D1629273

Andreas Vogel
Hubert Miller
Reinhard Greiling

The Rhenish Massif

earth evolution sciences

International Monograph Series
on Interdisciplinary
Earth Science Research and Applications

Editor
Andreas Vogel, Berlin

Andreas Vogel
Hubert Miller
Reinhard Greiling
(Eds.)

The
Rhenish Massif

Structure, Evolution, Mineral Deposits
and Present Geodynamics

Friedr. Vieweg & Sohn   Braunschweig/Wiesbaden

CIP-Kurztitelaufnahme der Deutschen Bibliothek

The **Rhenish Massif**: structure, evolution,
mineral deposits, and present geodynamics/
Andreas Vogel ... (eds.). — Braunschweig;
Wiesbaden: Vieweg, 1987.
   (Earth evolution sciences)
   ISBN 3-528-08967-9

NE: Vogel, Andreas [Hrsg.]

Produced by W. Langelüddecke, Braunschweig
Printed in Germany

ISBN 3-528-08967-9

Contents

Editorial

This book represents a selection of papers from an international and interdisciplinary workshop-meeting which was held in Boppard/Rhein in 1984.

Researchers from various fields of earth-sciences met to report on the results of their research, to discuss problems involved in present research activities, and to plan future work.

Necessarily a conference on this subject has to be international. The Rhenish Massif forms a geological unit with the Ardennes, which extend into western neighbor countries of Germany. In addition, we felt the necessity to use English as the conference language in order to provide the opportunity of active participation to those colleagues in foreign countries who are directly concerned with topics of the conference or interested in various aspects from the viewpoint of geological correlation, earth history, or research methodology.

Topics treated at the conference were paleogeographic reconstructions, plate tectonic concepts and Paleozoic stratigraphy, Variscan orogeny, deposits of economic minerals, geophysical data in relation to structural geology, the Eifel-Postvariscan evolution, as well as recent and present geodynamics.

The reader should however not expect a comprehensive presentation of the geological evolution of the Ardenno-Rhenish Massif. The contributions which are included are mainly original research papers on specific problems. In our present time, where the theories of global tectonics have created spectacular advancements in earth sciences, one sometimes feels the tendency that proposed hypotheses are supported on a very weak basis of fundamental facts. For further scientific progress we need working hypotheses, but also detailed investigations on basic material and careful collection and analysis of original data.

In order to meet requirements of readers who want to gain a wider view on the geological evolution and present evolutionary processes of the Rhenish Massif or who want to penetrate deeper into details, we have included extended abstracts with references to papers which have appeared elsewhere. One paper is about structural research on the Bohemian Massif, which is of great significance in comparison with the Rhenish Massif.

Finally we hope that the workshop-meeting and the outcome which we present here will contribute to stimulate further research on earth history of the Rhenish Massif, which is challenging both from the scientific standpoint and fascinating in view of the beautiful landscape created by both internal and external forces.

A. Vogel, H. Miller, and *R. Greiling*

Facies Development and Paleoecology at the Lower/Middle Devonian Boundary in the Southwestern Ebbe Anticline (Rheinisches Schiefergebirge) and Paleogeographic Interpretation

H. Avlar

Geologisch-Paläontologisches Institut der Westfälischen Wilhelms-Universität Münster, Correnstr. 24, 4400 Münster, Federal Republic of Germany

Key Words

Lithofacies
Biofacies
Paleoecology
Paleogeography
Ebbe Anticline

Abstract

The development of the Remscheid-, Cultrijugatus- and Hobräcke-Formations in the western Ebbe anticline (Meinerzhagen and neighboring sheets) is described by biological and lithological facies analysis.
Examination of three different facies types and paleoecological associations yields continual variations from the center of Meinerzhagen sheet towards the periphery.
Syn-sedimentary vertical movements have to be assumed, which led to the lateral facies differentiation. The so-called "Wilbringhäuser Querhorst" of previous workers therefore represents a positive area already active in Lower Devonian times. It is not a phenomenon tectonically set up.

Introduction

The worked area is part of the hilly landscape of the western Sauerland, situated at the southwestern flank of the Ebbe mountains between Gummersbach in the South, Lüdenscheid in the North, Wipperfürth in the West and Herscheid in the East (Fig. 1). The geological frame is set by the Ebbe anticline consisting of older Paleozoic rocks and Lower to Middle Devonian beds. The units under work are the Remscheid-formation (uppermost Upper Emsian), and the Cultrijugatus- and the Hobräcke-formations (lowermost Lower Eifelian).

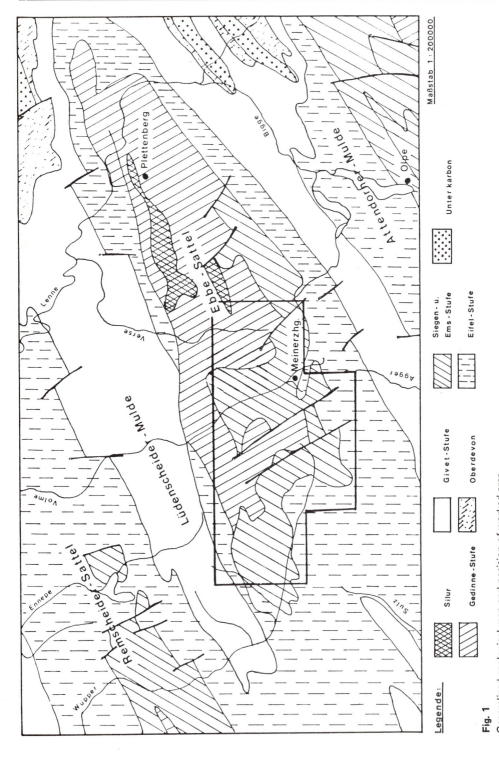

Maßstab 1 : 200000

Legende:

	Silur		Givet-Stufe		Siegen- u. Ems-Stufe
	Gedinne-Stufe		Oberdevon		Eifel-Stufe
					Unter karbon

Fig. 1

Generalized geological map and position of worked area

In the worked area, the Remscheid-formation is tripartite in Meinerzhagen sheet and Gummersbach sheet (Fig. 2). This stratigraphic division, however, is applicable only in the center of the region because of great facies variation.

The Remscheid-formation documents agitated water in the depositional environment, which is assumed to have been shoaly by many authors. Analysing the Remscheid-formation exactly at either low or large scale, surprising results may be found with respect to the facies. Three out of fifteen facies types are presented in the following sections.

Facies Types of the Remscheid-Formation

Lithofacies of the Remscheid-Formation

The Remscheid-formation occurs in 27 sheets in the Rheinisches Schiefergebirge varying widely in lithology: all transitions are found from shaly, finely laminated over silty, rough, partially sandbanded flaser schists to mica-rich, occasionally quarzitic sandstones. Freshly cut siltstones and slates are bluish-grey, rarely black; sandstones usually are grey or greenish-grey.

Out of three or four lithofacies encountered in the mapped area, only a characteristic one is described.

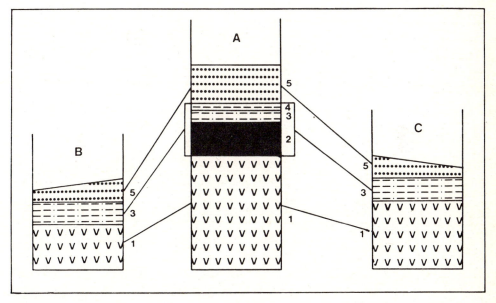

Fig. 2

Stratigraphic development of the lower Remscheid-formation. The blackshale is concordant with the Hauptkeratophyr. Note profiles A, B, C as marks for the "bluish-grey shales" developed in other sheets:

A: Meinerzhagen sheet near Wiebelsaat
B: Wipperfürth sheet south of Kerspe dam
C: Herscheid sheet near Hösinghausen

1 Hauptkeratophyr, **2** blackshale, **3** silty shale, **4** brownish-yellow-greenish silty shale, **5** feldspathic sandstone

Blackshale Facies

The blackshale is a typical euxinic basinal sediment. It is a dark grey, mostly black, laminated shale rich in sulfides and containing organic matter and fossils. The latter derive from swell or coastal regions and were transported into the basin of blackshale sedimentation postmortally.

Limitation and Thickness: The blackshale is concordant with the Hauptkeratophyr (K 4). A reddish brown siltstone forms the capping bed. According to Fuchs (1923), the blackshale was put into the lower Wiebelsaat-formation as "greyish-blue shales".

The Wiebelsaat-formation forms the facies analogue of the lower Remscheid-formation in the mapped area. That is why I will designate it as the lowest part of the lower Remscheid-formation.

The greatest thickness of blackshale is achieved between the townships of Vorth and Wiebelsaat (Meinerzhagen sheet); there it varies from 4 to 5 meters.

Horizontal Limitation: Fuchs (1915) recorded the complete Wiebelsaat-formation on the western and southern slope of the Ebbe anticline (Wipperfürth, Meinerzhagen and Herscheid sheets) as a well-distinguished horizon.

Fuchs (1923) restricted the blackshale facies to the Meinerzhagen sheet. In the Herscheid sheet it was equated by him as "bluish-grey shales" with the type of the Remscheid-formation.

As a special facies the blackshale is situated mainly between Werfelscheid and Sulenbecken north of Meinerzhagen. It is wedging out in the Herscheid sheet in the east and in the Wipperfürth sheet in the west (Fig. 3). In the Gummersbach sheet it could be encountered thinly between Rathlendorf and Börlinghausen.

Origin of the Blackshale: Blackshales are deposited in environments strongly depleted in oxygen. Due to missing or weak bioturbation, the lamination is well-preserved (Wetzel 1982).

Further investigations, however, show that blackshales in geologic history were formed in very shallow open oceanic realms (especially in the Central European Variscan Belt), e.g. in closed partial basins, on the shelf, between submarine volcanoes, or between growing reefs (Krebs 1969). This environment of blackshales may be situated nearshore on the shelf edge (Einsele & Wiedmann 1982; Thurow, Kuhnt & Wiedmann 1982).

Wiedmann (1982) explained the genesis of blackshales by a subsidence model with the development of the passive continental edge of Central Morocco as an example. The evolution took place after an early rifting- and evaporite-state accompanied by little depth of lowering and lacking circulation.

Areal occurrences of blackshales are said to be either distal turbidites or sediments of local upwelling depending on submarine or coastal relief. As Thurow, Kuhnt & Wiedmann (1982) point out by analysis of Turonian blackshales of the Atlas, these upwelling sediments owe their character to their situation in passat zones.

Eustatic changes of the sea level, especially transgressions, are closely connected to the origin of blackshales throughout geologic history (Demaison & Moore 1980; Krebs 1969). During transgressions, coastal areas were eroded because of the rising sea level. The generated erosional debris released a great amount of nutrients in the water on the one hand; on the other hand, the basins finally lay below wave base due to a further rising sea level. Thus euxinic conditions were established that led to the deposition of blackshales (Paul 1982).

Vail et al. (1977, 1979) calculated a rise of the sea level of 120 to 270 meters above the present level in the course of the Devonian period. As a result the terrestrial facial character ended with the Hauptkeratophyr (K 4) and the transgressive phase quickly established

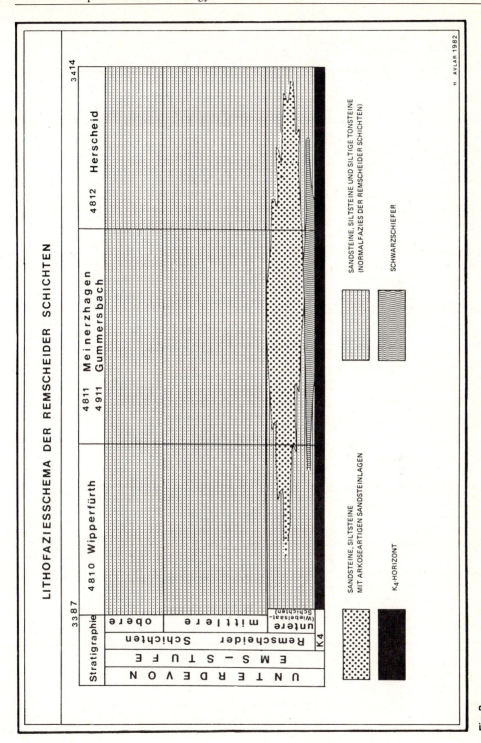

Fig. 3
Horizontal limitation and distribution of blackshale facies

fully marine conditions after the Remscheid-formation (Wong & Degens 1980). The sedimentation of blackshales was thus made possible by this transgression at the base of the Remscheid-formation, concentrating on the basin of Kierspe-Meinerzhagen.

Fig. 4 elucidates the lowering of the Kierspe-Meinerzhagen basin below the wave base during a rise of the sea level. Thus an euxinic environment originated facilitating the deposition of blackshale in this place. Since the Rheinisches Schiefergebirge was part of the Baltic Plate in Emsian times, situated $10-15°$ south of the equator (Ziegler 1981), an explanation of the blackshales as sediments of coastal upwelling (Einsele & Wiedmann 1982) could be possible as well because of the location in passat zones (Fig. 5).

The blackshale deposition stops with further rise of the sea level, because of which the Kierspe-Meinerzhagen basin was no longer isolated from the rest of the Devonian sea.

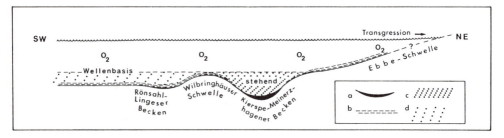

Fig. 4
Model of the origin of the blackshale in a closed partial basin under euxinic conditions.
a = blackshale, b = silty shale, c = stagnant water realm, d = presumably weakly agitated water

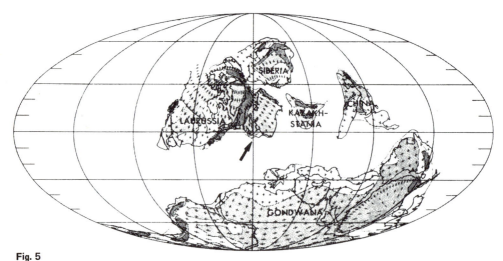

Fig. 5
Paleogeography of the Emsian (**Lower Devonian**); arrow indicates position of the Rheinisches Schiefergebirge at the southern border of the Baltic Plate (from Ziegler 1981)

Biofacies of the Remscheid-Formation

The fossil occurrences were examined in the whole district of Meinerzhagen, Gummersbach, Herscheid and Wipperfürth sheets. The graphical illustration of their abundance reveals a maximum in the Gummersbach and Meinerzhagen sheets (Fig. 6). Due to this rich content of fossils, the stress was put on biofacial analysis.

Coquina Facies

Layers of coquinas appear throughout the mapped area. In certain regions they are accumulated in huge masses, so they have to be described as independent facies.

From stratigraphic occurrence and sedimentary structures of their own and of the surrounding rocks, comparisons with Recent coquinas are made, which explain their origin and paleogeographic setting.

Limitation and Thickness: The stratigraphically lowest coquinas occur at the base of the middle Remscheid-formation. They reach up to the top of the lower Cultrijugatus-formation, increasing in abundance and thickness. Frequently, coquinoid layers are laterally interrupted. The total thickness of coquina layers and intercalated sediments is about 600 meters in the Kierspe-Meinerzhagen basin.

Horizontal Limitation: The shell bed-generating fossils occur in the whole area under study but do not everywhere form coquinas. The layers are concentrated in Kierspe, Meinerzhagen, Rönsahl, Marienheide, Lingese, Börlinghausen and Genkel, reaching their maximum abundance in the Kierspe-Meinerzhagen district (Fig. 7).

Origin of the Coquina Facies: In shallow seas, wind-moved water is responsible for transport and accumulation of organic skeletons (Schäfer 1962). Since tidal current and bottom-touching seaway depths of 100 meters may be achieved in the open shoaly sea (Schäfer 1941, 1962), the bottom-touching seaway transports every movable shell in the direction of the momentary current. Only nearshore is the migration of the shell interrupted and trash lines are formed together with other hardparts. Gullies are the only possibility to stop the shoreward transport before accumulation in trash lines.

The main migration paths of the shells take their course in the tidal zone (wadden sea) in the gullies and channels. Fig. 8 (after Krause 1950 and Schäfer 1962) illustrates the orientation of the paths from the wadden sea into the open ocean or vice versa, as is the present situation between the isles of Juist and Norderney (North Sea).

Schäfer (1962) compiled the sites of shell concentration in channels of the North Sea waddens between 10 and 15 meters depth. It may be distinguished between those "which (1) came down by the gullies, which (2) directly fell into the deep channels coming from the wadden sea, which (3) are kept in flutes on their migration along the shore, and which (4) accumulated from the open sea with flood current and seaway. The bottoms of the channels are divided from each other by plates or moving reefs and mostly lie deeper than the surrounding and even deeper than the bottoms of the sea extended in front; simply for this reason they have to get locations of shell trapping." (translated from German)

Summing up the characteristics of the coquina layers in the mapped area, the following can be said: epifaunal and infaunal gastropods, bivalves, brachiopods and other shell-bearing marine organisms of the Remscheid-time were either deposited loosely on the seafloor or stuck in the sediment at some centimeters depth. The shells and other hardparts were reworked to coquinas by the bottom-touching seaway or the tidal currents. After comparison with Recent examples, it has to be assumed that a great portion of shells was destroyed by early diagenetic processes.

10

H. Avlar

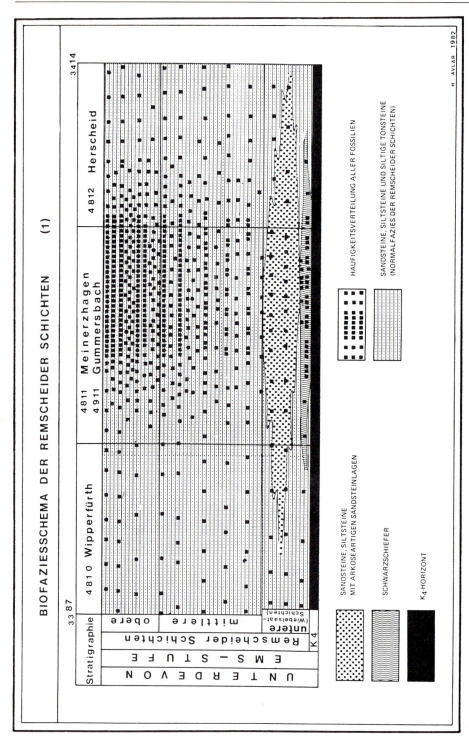

Fig. 6

Frequency distribution of fossil occurrences in the whole area

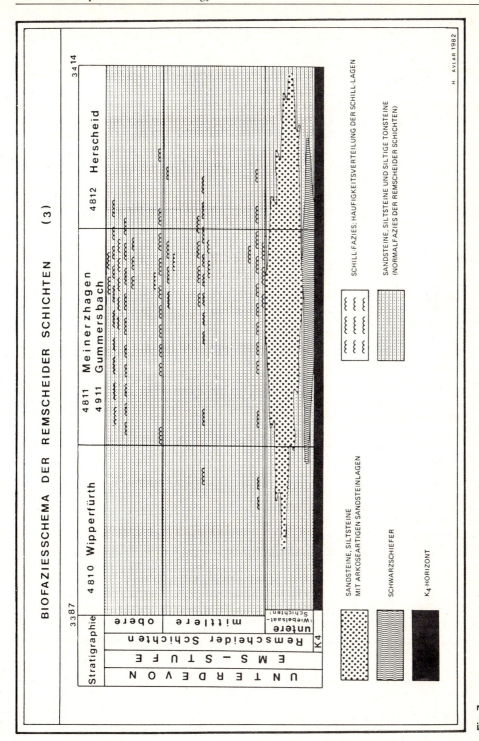

Fig. 7
Frequency distribution of coquina layers in the whole area

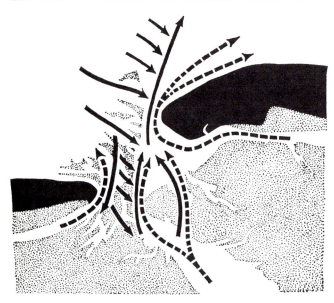

Fig. 8
Recent main migration paths of shells between the isles of Juist
and Norderney (North Sea):
black and white arrows = wadden coquina; black arrows = sea
coquina (after Krause 1950)

In spite of the important early shell dissolution, fossil coquina layers exist in huge masses
in the area under study; their thickness may reach some meters in the Remscheid-formation. In the lower part of the middle Cultrijugatus-formation most coquina layers are
located near Weh, Meinerzhagen and Kierspe-Blankenberg where they achieve a thickness
of more than 10 meters. These abundant fossil coquinas have to be deposited in similar
sites as in the Recent example Seegatt; that is, in gullies, channels and flutes.
The main concentration of coquinas in the area is in the surrounding of the townships
Kierspe and Meinerzhagen. In the neighborhood of the Wilbringhausen swell-complex,
migration paths of shells (Channels) can be stated. Fig. 9 shows the course of the most
important coquinoid beds in the district mentioned above.
The weak tectonical deformation facilitates the recognition of this system of gullies
and channels in the ancient wadden sea.

Plant Facies

The plant facies consists of autochthonous and allochthonous plant remains and a few
underclay horizons. In the mapped area, plant fragments are so abundant that they
build up a welldefined facies type comprising underclays, which were encountered in
four beds (as known so far).
The plant remains were analysed as follows: their relative abundance was stated in the
whole area, the length of fragments was measured in the field, and underclays were
mapped. As a result, it can be stated that the fragments are concentrated in locations
where underclays exist; the longest plant remains (25—30 cm) were found near Wiebelsaat.

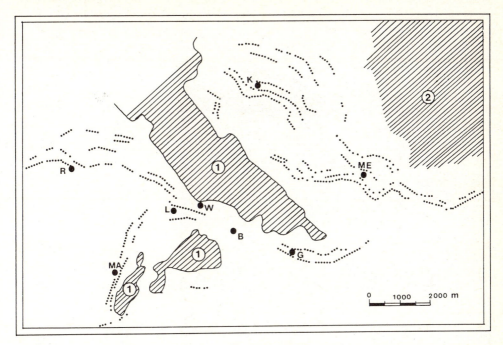

Fig. 9
Channel systems of Remscheid- and Cultrijugatus-formations as drawn from shells' migration paths.
1 = Wilbringhausen swell-complex, **2** = Ebbe swell, **M** = Meinerzhagen, **K** = Kierspe, **G** = Genkel, **L** = Lingese, **B** = Börlinghausen, **R** = Rönsahl, **M** = Marienheide

Limitation and Thickness: Plant fragments occur in the whole Remscheid-formation. They are accumulated in the lower and middle Remscheid-formation where underclays exist as well.

Horizontal Limitation: a) Underclays: Underclays were observed once in the Herscheid sheet and frequently in the Meinerzhagen sheet. Here they occur at Ohl-Singerbring and Auf dem Mark (lower Remscheid-formation), Neu-Grünthal in two horizons and Immeke (middle Remscheid-formation). In the Herscheid sheet, they crop out near Valbert (middle Remscheid-formation).

The horizons with underclays were pursued into the neighboring sheets, but either these beds were not deposited or they do not contain underclays.

b) Plant remains: The abundance of plant fragments decreases from the center of the worked area towards the East and the West. Statistical evalution revealed a correspondence of lengths of plant remains with the abundance; lower values are achieved in the East and in the West (Fig. 10).

Origin of the Plant Facies: The first underclays in the Remscheid-formation of the mapped area were discovered by Drees (1971) in the Herscheid sheet. Homann (1980) found underclays in two different outcrops near Ohl-Singerbring and Neue Brücke (Meinerzhagen sheet). They were classified as nonmarine sediments by him (pers. comm. Remy and Ortlam) and regarded as autochthonous because of the following observations:
— high angle of ramification, as characteristic of land plants
— structural preservation and hints toward a bark

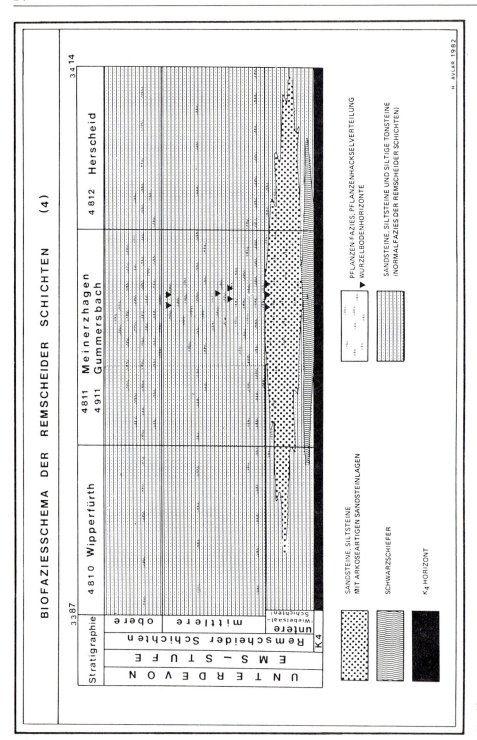

Fig. 10
Distribution of plant fragments and position of underclay horizons in the whole area

— sporangia
— hints of soil development at the top of the profile
— precipitation haloes around the imbedded plants, which can only exist because of "bleeding" of plants covered with sediment directly after their death
— increasing lack of structure in the sediment from base to top
— indistinct lower boundary
— sharp upper boundary
— branched tubes of roots reaching up to 10 cm length

Such autochthonous or hypautochthonous plant beds and underclays can only form above the sea-water level (Homann 1980). All underclays of the Lower and Middle Devonian belong to the permanently moist to wet soils (groundwater-, stagnant water- or submerged soils) (Remy 1980).

In the outcrop of Neubrücken, two underclays were ascertained in a vertical distance of 8—10 meters (confirmed by pers. comm. of Ortlam 1982 in Münster). Above and below such horizons, marine fossil layers were frequently encountered (Fig. 11); so the plants were embedded autochthonously by an inundation. Below the plants a clayey bed of 10 to 15 centimeters is rooted. The plants were covered by silty and sandy sediments.

In these tempestite-like deposits brachiopods, tentaculites and other marine fossils are found.

Starting with the Hauptkeratophyr (K 4), the intensified transgressive phase quickly brought the Remscheid-formation into brackish and finally fully marine conditions (Wong & Degens 1982). According to Vail et al. (1977, 1979), the sea level permanently rose since the base of the Devonian. This rise did not take place continuously, but was interrupted by regressive phases several times. The oscillations of the sea level originated isles and wadden areas; that is, the Wilbringhausen swell-complex frequently existed as an isle with surrounding tidal flats.

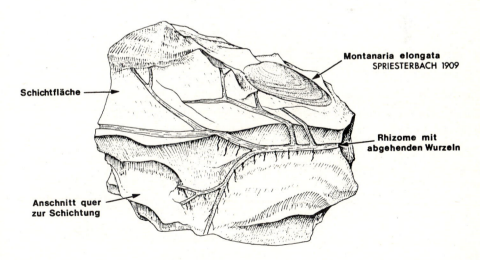

Montanaria elongata
SPRIESTERBACH 1909

Schichtfläche

Rhizome mit
abgehenden Wurzeln

Anschnitt quer
zur Schichtung

Taeniocrada sp.

Fig. 11
Underclay with marine bivalve (drawn from nature)

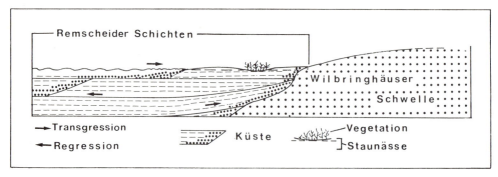

Fig. 12
Schematic sketch of spatial and temporal surrounding of plant biotopes in the Remscheid-formation

Since the Rheinisches Schiefergebirge was a southern part of the Baltic Plate in Emsian Times — thus positioned about $10-15°$ south of the equator (Ziegler 1981) — the assumption of a moist and warm climate seems reasonable. Together with the paleogeographic conditions described above, the circumstances were ideal for the development of higher plants. The following scenario may be sketched:

a) The sediments of the Remscheid-formation were drained during a regressive phase.
b) In the course of transgressions the wadden areas formed, and in the nearshore lowland plains the groundwater level rose. The foreshore of the isles with lagoons and swamps offered a good environment for the growth of plants (Fig. 12).
c) The sustained rise of the sea level caused the sedimentary cover of the lowland plants, but at the same time new possibilities of growth were created in higher coastal regions.
d) The plants not directly embedded during an inundation were fragmentarily accumulated in trash lines together with marine biota; thus the beds rich in plant fragments of the Remscheid-formation originated.
e) These processes were repeated at least four times in the course of the deposition of the Remscheid-formation, which resulted in a mixing of marine fossils and sediments with horizons containing allochthonous, hypautochthonous and autochthonous plants.

By permanent inundations, the vegetation was repeatedly destroyed. The flora colonizing shorelines and tidal flats was continuously influenced by the sea water.

Paleogeography

Out of 15 worked facies types, only three were outlined briefly. The lateral differentiations within each single facies contribute to a joint picture including paleoecology and biotopal evolution as well. Therefrom secure paleogeographic statements can be made:

It was necessary to reinterprete an area in which Lower Devonian beds are situated within younger sediments (uppermost Upper Emsian and Lower Eifelian). These beds being surrounded by the Remscheid-formation. In the geological map mostly belong to the Gedinnian, occuring in the Meinerzhagen sheet and the northern part of the Gummersbach sheet. The Gedinnian age is verified by fossils.

Fuchs (1915) interpreted the beds as a tectonical phenomenon and called them "Querhorst von Wilbringhausen", later (1923) "Wilbringhäuser Querhorst". This theory was principally accepted by later workers (Spriestersbach 1925; Schriel 1936a, b), until Wiegel (1960) reinterpreted the region.

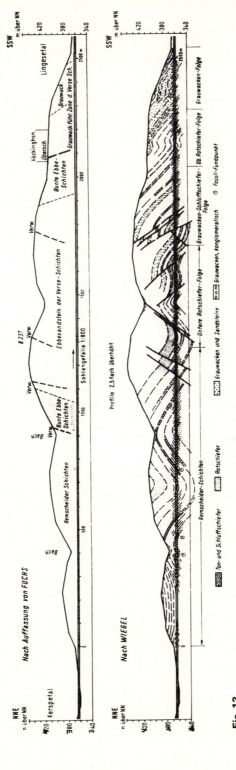

Fig. 13

Profile cutting through the northeastern part of the Wilbringhausen main swell with transition zone to the Remscheid-formation. above: opinion of Fuchs; below: opinion of Wiegel (from Wiegel 1960)

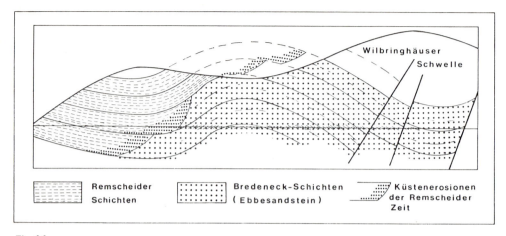

Fig. 14

Weakly deformed transition zone of Remscheid-formation and Wilbringhausen swell with Gedinnian beds

He analysed a profile from Kierspe to Lingese in the run of the drift of a gallery across the "horst". Not having found a cross fault fitting to the character of such a tectonic unit, he assumed special folding of this area within the Remscheid-formation and called it "Wilbringhaus block" (Fig. 13).

My own research yielded the existence of Gedinnian beds (in contrast to Wiegel 1960) and the non-existence of major tectonic faults (in contrast to Fuchs 1913) (Fig. 14). I analysed temporary outcrops of a waterpipe-lining in the horst area and, similarly to Wiegel (1960), seized a transitional region from the horst into the Mühlenberg-Sandstone (Eifelian) void of important faults.

Paleoecological and facial results render a new paleogeographic interpretation: the Gedinnian beds already in the upper Lower Devonian (Emsian) partially formed a positive area with lowered rates of subsidence amidst the geosyncline. In its surrounding, basins were formed because of relatively stronger depression (Kierspe-Meinerzhagen, Rönsahl, Marienheide, Börlinghausen and Genkel). Such differentiated movements can be traced in the Remscheid-formation, the Cultrijugatus-formation, and the Hobräcke-formation. During this time interval, an isle existed in the area of the Gedinnian beds sinking gradually afterwards in common with the surrounding. A littoral facies can be substantiated in the Remscheid-formation, and reef formation occurs at the boundary of the Cultrijugatus-formation and Hobräcke-formation due to the now initiated lowering.

The Wilbringhausen swell-complex may be subdivided: in addition to its main range in the Meinerzhagen sheet, two further blocks exist southwesterly in the Gummersbach sheet. Even the main range is made up of a northeastern and a southwestern part. In between, a smaller basinal area is situated (Kierspe-Meinerzhagen and Lingese); here partially (?Siegenian and) Lower Emsian beds crop out, which are not fully eroded and became folded in common with the existing Gedinnian. As already Oncken (1982a, b, 1984) pointed out, all major structures of the northern Rheinisches Schiefergebirge owe their character to a congruent pattern of synsedimentary lowering and differential uplift. The results displayed rest on facial and paleoecological examinations, but the distribution of swells and basins is proved by the graph of thicknesses. Fig. 15 shows their three-dimensional depiction in a block diagram based on reciprocal values of bed thickness. On the one hand, regions can be seen with low thicknesses because of low sedimentation rates acting as shoals or positive areas during Remscheid- and Cultrijugatus-times (1, 2 in Fig. 15); on the other hand, basins show up, e.g. the Kierspe-Meinerz-hagen basin (K, M in Fig. 15).

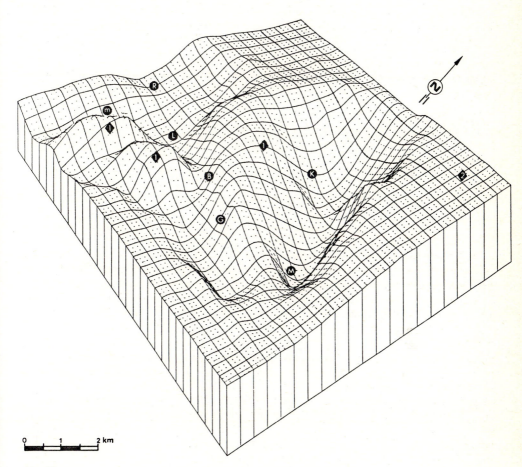

0   1   2 km

Fig. 15
Block diagram of worked area based on bed thickness from Hauptkeratophyr up to Cultrijugatus-formation. Note locally different subsidence rates within this time interval (strongly exaggerated vertically)
Areas of low subsidence and sedimentation rate

1	Wilbringhausen swell complex	2	Ebbe-swell

Basinal areas of high subsidence and sedimentations rate

B	Börlinghausen	M	Meinerzhagen
G	Genkel	m	Marienheide
K	Kierspe	R	Rönsahl
L	Lingese		

References

Demaison, G. J. and G. T. Moore (1980): Anoxic environments oil source bed genesis. Mem amer. Assoc. Petrol. Geol. 64, 1179—1209.

Drees, H. (1971): Stratigraphische und tektonische Untersuchungen am Südrand des Ebbegebirges an der Grenze Unter-Mittel-Devon im Raume Meinerzhagen-Valbert. Diss. Univ. Aachen, 143 S.

Einsele, G. and J. Wiedmann (1982): Turonian black shales in the Moroccan coastal basins: First upwelling in the Atlantic Ocean. — In: v. Rad, U. et al. (Eds.): Geology of the Northwest African Continental Margin. Springer, Berlin, Heidelberg, New York.

Fuchs, A. (1915): Die Entwicklung der devonischen Schichten im westlichen Teil des Remscheid-Altenaer-Sattels und Ebbe-Sattels. Jb. preuss. Landesanst. 36, 1—95.

Fuchs, A. (1923a): Erläuterungen zur geologischen Karte von Preußen und benachbarten Bundesstaaten. — Liefg. 220 MTB 4811 Meinerzhagen. K. preuss. Landesanst., 52 S.

Homann, H. (1980): Geologie und geochemische Charakteristik des höheren Unterdevons im Bereich von Wiebelsaat, östlich Kierspe. Dipl.-Arb. Univ. Münster, 117 S. (unpubl.).

Krause, H. R. (1950): Quantitative Schilluntersuchungen im See- und Wattengebiet von Norderney und Juist und ihre Verwendung zur Klärung hydrographischer Fragen. Arch. Molluskenkde 79, 91—116.

Krebs, W. (1969): Über Schwarzschiefer und bituminöse Kalke im mitteleuropäischen Variscikum. Erdöl und Kohle 22, 2—6.

Oncken, O. (1982a): Zur Rekonstruktion der Geosynklinalgeschichte mit Hilfe von Inkohlungskurven (am Beispiel Ebbeantiklinorium, Rheinisches Schiefergebirge). Geol. Rdsch. 71, 579—601.

Oncken, O. (1982b): Determinierung und Entwicklung großtektonischer Strukturen im nördlichen Rhenoherzynikum (Beispiel Ebbeantiklinorium). Diss. Univ. Köln, 198 S.

Oncken, O. (1984): Zusammenhänge in der Strukturgenese des Rheinischen Schiefergebirges. Geol. Rdsch. 74, 619—649.

Paul, J. (1982): Zur Rand- und Schwellen-Fazies des Kupferschiefers. Z. dtsch. geol. Ges. 133, 571—605.

Pitman, W. C. (1979): The Effect of Eustatic Sea Level Changes on stratigraphic Sequences at Atlantic Margins. Mem amer. Assoc. Petrol. Geol. 29, 453—460.

Remy, W. (1980a): Wechselwirkungen von Vegetation und Böden im Paläophytikum. Festschrift Gerhard Keller, 43—73.

Schäfer, W. (1941a): Zur Fazieskunde des deutschen Wattenmeeres. 1. Dangast und die Ufersäume des Jadebusens. Abh. senckenb. naturforsch. Ges. 457, 1—33.

Schäfer, W. (1941b): Zur Fazieskunde des deutschen Wattenmeeres. 2. Mellum, eine Düneninsel der deutschen Nordsee-Küste. Abh. senckenb. naturforsch. Ges. 457, 34—54.

Schäfer, W. (1962): Aktuo-Paläontologie nach Studien in der Nordsee. 666 S., Frankfurt a. M.

Schriel, W. (1935): Siegener und Koblenzschichten in der Grauwacken führenden Zone und dem Ebbesandstein des westlichen Ebbesattels. Z. dtsch. geol. Ges. 87, 40—47.

Schriel, W. (1936): Das Unterdevon im südlichen Sauerlande und Oberbergischen. Festschr. z. 60. Geb. Hans Stilles, 1—21.

Spriestersbach, J. (1925): Die Oberkoblenzschichten des Bergischen Landes und Sauerlandes. Jb. preuss. geol. Landesanst. XLV, 367—450.

Thurow, J., W. Kuhnt, and J. Wiedmann (1982): Zeitlicher und paläogeographischer Rahmen der Phthanit und Black Shale-Sedimentation in Marokko. N. Jb. Geol. Paläont. Abh. 165 (1), 147—176.

Vail, P. R. and R. M. Jr. Mitchum (1979): Global Cycles of Relative Changes of Sea Level from Seismic Stratigraphy. Mem. amer. Assoc. Petrol. Geol. 29, 469—472.

Vail, P. R., R. M. Jr. Mitchum, and S. Thompson (1977): Seismic stratigraphy and global changes of sea level, Part 4: global cycles of sea level. — In: Payton, C. E. (Ed.): "Seismic Stratigraphy-Application to Hydrocarbon Exploration". Mem. amer. Assoc. Petrol. Geol. 26, 83—97.

Wetzel, A. (1982): Zeitliche Klassifikation von Sedimentationsprozessen bei der Schwarzschiefer-Bildung. N. Jb. Geol. Paläont. Abh. 165 (1/2), 30—31.

Wiedmann, J. (1982): Grundzüge der kretazischen Subsidenz-Entwicklung im Südatlantik, in Marokko, Nordspanien und im Helvetikum. N. Jb. Geol. Paläont. Abh. 165 (1), 5—31.

Wong, H. K. and E. Degens (1980): Geotektonische Entwicklung des variszischen Faltungsgürtels im Paläozoikum. Mitt. Geol. Paläont. Inst. Univ. Hamburg 50, 17—44.

Ziegler, A. M. (1981): Paleozoic paleogeography. — Paleoreconstruction of the continents. Geodyn. Ser. 2, 27—37.

Late Hercynian Plate and Intraplate Processes within Europe

V. Lorenz

Institut für Geowissenschaften, Universität Mainz, Saarstr. 21, D-6500 Mainz, Federal Republic of Germany

The Hercynian orogenic belt of Europe consists of a central crystalline ridge which is accompanied on both sides by a rather unmetamorphosed foldbelt. It is speculated that the crystalline ridge represents some kind of island arc system underlain by a segment of continental crust. On both sides this island arc system was involved in subduction of oceanic crust, first of the Mideuropean Sea in the North and then of the Paleotethys in the South. When the continental areas to the north and south of the oceanic areas (North America/Northern Europe and Africa) finally got involved in the subduction processes, continent/continent collision took place on both sides of the island arc system. The two subduction zones encroached, from the Visean onwards, onto the shelf areas of the two large continental plates. As a consequence the Rhenohercynian foldbelt formed on the shelf area of North America/Northern Europe (E Sudetic Mts., Harz, Rheinische Schiefergebirge, SW England, S Portugal, and S Appalachians) whereas in the South the foldbelt formed on the African shelf along the Carnian Alps, the Montagne Noire, northern Iberia, and finally southeastern Morocco and the eastern Mauritanides.

Because of the irregular continental margins of the large, thick continental plates approaching each other, the island arc system of Hercynian Europe was sandwiched in between, and it became deformed plastically by oroclinal bending because it was a long, narrow and hot plate of reduced rigidity.

During the two continent/continent collisions, the two areas of compressive deformation from the Visean onwards were separated by an increasingly large area of crustal uplift and crustal extension. An area of basin and range topography formed. After the collision processes had ended, uplift continued and even more extensive crustal extension took place lasting for several tens of millions of years. Thus about 70 intermontane basins formed within Hercynian Europe and the neighboring area of Northern Europe during uppermost Carboniferous and lowermost Permian time. Within nearly all these intermontane basins, thick clastic continental sediments and large amounts of volcanic rocks accumulated. Waning of the geodynamic activity within the Hercynian orogenic belt caused crustal subsidence and formation of a peneplain in Permian time. Finally in Upper Permian time, transgression of both the Zechstein Sea in the North and the Tethys in the South marked the end of the Hercynian geodynamic cycle.

References

Lorenz, V. and Nicholls, I. A. (1984): Plate and intraplate processes of Hercynian Europe during the Late Paleozoic. Tectonophysics, 107, 25–56.

Event-Stratigraphy of the Belgian Famennian (Uppermost Devonian, Ardennes Shelf)

R. J. M. Dreesen*

Alexander von Humboldt Fellow, Geologisches Institut der RWTH-Aachen, Wüllnerstr. 2, D-5100 Aachen, Federal Republic of Germany

Key Words

Upper Devonian
Biotic crises
Eustatic changes
Non-deposition
Oolitic ironstones
Synsedimentary tectonics
Seismics and volcanism

Abstract

The presence of a condensation near the Frasnian-Famennian boundary most probably results from an eustatic sea-level fall and subsequent non-deposition events on the Ardennes shelf.
The episodic occurrence of oolitic ironstones on the shelf and their synchronism with turbidites and volcanic deposits within the more pelagic settings of the Rheinisches Schiefergebirge is not a coincidence, but rather the result of interdependent tectono-sedimentary processes. These excellent event-stratigraphical marker beds are the mute evidences of turbulent episodic events related to pre-orogenic synsedimentary tectonic movements within the Ardenno-Rhenish Massif during late Devonian times.

Introduction

Event Stratification

Event stratigraphy — the new fashion of the eighties?
During this last decade a growing number of scientific papers and symposia are being devoted to the importance of rare or "abnormal" geological events, on their possible causes, and on their influence on evolution, ecology and stratigraphy.

---

* actual address: Institut National des Industries Extractives (INIEX), Rue du Chéra 200, B-4000 Liège, Belgium

The present generation of geoscientists proclaims that sudden and violent processes (= rare or episodic events) are not exceptional geologically and that these are to be explained as perfectly natural rather than supernatural.

Despite the fact that uniformitarianism has influenced our geological thinking for more than 150 years, it has become clear now that the sedimentary record is largely a record of episodic events rather than being uniformily continuous: episodicity is the rule not the exception! (Dott, 1982)

In contradiction with cyclic sequences (periodites) — which are due to slow, gradual periodic changes in sediment parameters, bioturbation and sedimentation rates with time — episodic events are the result of abrupt changes in those parameters and in the build up/time curve (Einsele & Seilacher, 1982).

The gradual changes of periodites seem to be related to the earth's orbital cycles of precession, obliquicity and eccentricity, which affected the climate on the continents, the current systems of the oceans, the sea level and some physicochemical properties of the ocean waters.

On the other hand episodic or rare events — rare in terms of human life spans and experiences — represent larger deviations from normal intensity: these include events of greater-than-normal intensity such as volcanic and seismic activity, turbidity currents, severe storms, flood deposits, and events of less-than-normal intensity such as non-deposits, and mineralized hardgrounds. Both types may produce event depositions or surfaces, all of which represent short episodes on the geological time scale (Dott, 1982).

Since turbulent events do last only for relatively short time intervals, they represent high-resolution tools in stratigraphic correlation and basin analysis.

Biotic Crises — Global Events

Extremely violent processes such as tsunami could have done a great deal of the sedimentary work but we have not yet learned to recognize their results. Yet even more powerful waves generated by very large meteorite or asteroid impacts in the ocean may be important. These "catastrophic" events are receiving much interest now as a result of the speculation that such an impact might have caused global biotic crises (extinction events) such as the Cretaceous-Tertiary and the Frasnian-Famennian boundary extinctions (McLaren, 1982, 1983). These extinctions appear to be selective and they apparently affected essentially the marine plankton and the shallow-water benthos throughout the tropic and subtropic regions of the world.

In the marine realm, the immediate causes of such abrupt or "bedding-plane" extinctions might be a worldwide poisoning of the oceans by mixing of anoxic waters (black shale deposition events), an increase in suspended sediment by prolonged turbidity, rapid regressions or sudden temperature changes (by melting of pack-ice, Copper, 1984). Several of these, in turn, might be caused by a large body impact or a long-term volcanic activity. In the terrestrial realm, climatic stress, temperature changes and effects caused by large body impacts or major and prolonged volcanic activity (dust leading to darkness in order of several months duration and high N-oxide-concentration in the atmosphere) are believed to be the major offenders. Volcanic injections of CO_2 (perhaps accompanied by Iridium) into the atmosphere would have had substantial effects on biota, as would other catastrophic events like tsunami and floods due to (submarine) volcanic eruptions.

Besides the fact that such global events are characterized by a huge extinction of biomass and major disappearances of taxa, the importance of those extinction horizons lies in the potential for world-wide accurate stratigraphic correlation. These bedding-plane extinctions events are the kind of boundaries we must look for, instead of using "quiet" boundaries selected more or less arbitrarily for their convenience (as for instance the Silurian-Devonian boundary) (McLaren, 1984)

Famennian Event-Stratigraphy

In this paper it is demonstrated that episodic events influenced the depositional history of the shelf, south and southeast of the London-Brabant Massif during late Devonian times (Fig. 1).

Particular events may also represent excellent stratigraphical marker beds throughout the different tectonic units of a same sedimentary basin. These event-stratigraphical marker beds often form isochronous horizons by which widely-spaced and distinct sedimentary facies can be correlated. However, only a multidisciplinary approach can result in the recognition of true event-stratigraphical marker beds, as they are the result of strongly interdependent processes.

It is suggested here that among the tectono-sedimentary mechanisms which controlled the paleogeographical evolution of the Rhenohercynian Zone (Fig. 2), synsedimentary tectonic movements along deep-seated faults or structural lineaments (transversal to the variscan strike, Dvorak, 1973) must have been very important. The presence of such faults deep in the Paleozoic basement is indirectly indicated by varying thicknesses of sediments, flexures, abrupt change in facies, and by the distribution of volcanic centres and reefs (from Lower Devonian trough early Carboniferous times).

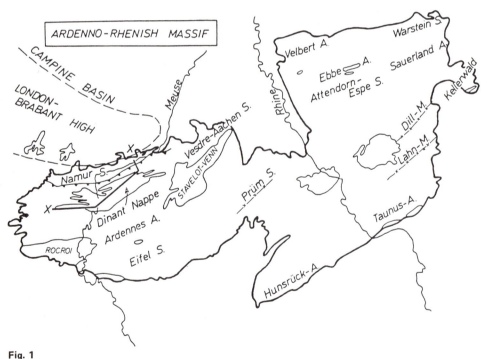

Fig. 1
Tectonic sketch map of the Ardenno-Rhenish Massif. Line XX refers to cross section of Fig. 2.

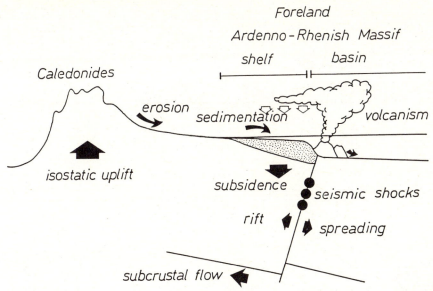

Fig. 2

Cartoon showing the different tectono-sedimentary processes affecting the Rhenohercynian Zone (after Walliser, 1980).

An important regression has controlled the Famennian depositional environment of the Belgian Ardennes (neritic shelf). The paleogeographical evolution of this shelf has been reconstructed on the basis of natural outcrops and well data from different tectonic units: the Namur, Dinant and Vesdre Synclinoria (Fig. 1).

This regression started at the end of the Frasnian and reached its acme with the deposition of the so-called Condroz Sandstones. This regressive megasequence is well-illustrated in the Dinant Nappe by the prograding movements of longshore sand barriers, closely followed by the development of intertidal, tidal lagoonal, coastal sabkha and even alluvio-lagoonal environments (Fig. 3; Thorez et al. 1977; Dreesen & Thorez, 1980, 1982).

The Frasnian-Famennian Boundary Event

A first important event occurs near the Frasnian-Famennian stages transition. Here a sudden sea-level fluctuation provoked a non-deposition event on the Ardennes shelf while a particular facies — the bituminous black shales and Kellwasser limestones — developed in the more pelagic-hemipelagic settings of the Rheinisches Schiefergebirge (Fig. 5).

These Kellwasser facies and other related facies such as the Annulata shales (Upper Famennian) and the Hangenbergschiefer or "Liegende Alaunschiefer" (Dinantian) must be regarded as isochronous markerbeds, according to actual biostratigraphic data. They are not restricted to any particular area of the Rhenohercynian Zone, but occur widespread within the European Variscides and even worldwide. Therefore these kind of black shale events are most probably the result of global events which had an important impact on the ecological and depositional conditions. Krebs (1969, 1979) noticed already that

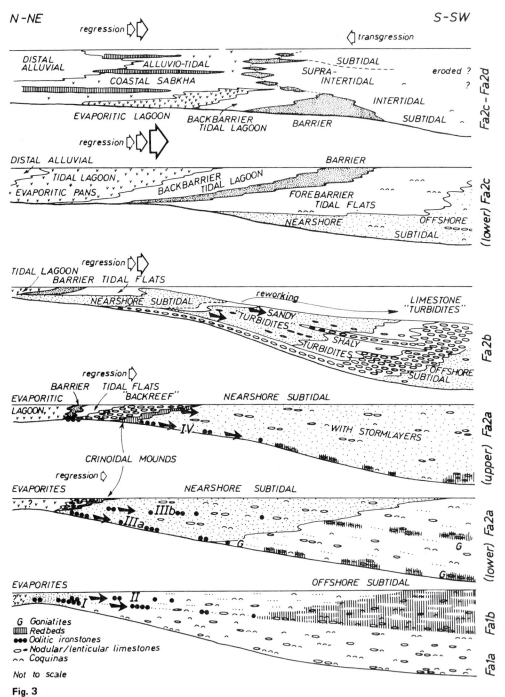

Fig. 3
Famennian regressive megasequence on the Ardennes shelf. NNE-SSW-cross section through the Dinant Nappe. Paleogeographical evolution at different time intervals (Belgian stratigraphic units). Lower (Fa 1) through Upper (Fa 2) Famennian (after Dreesen & Thorez, 1982).

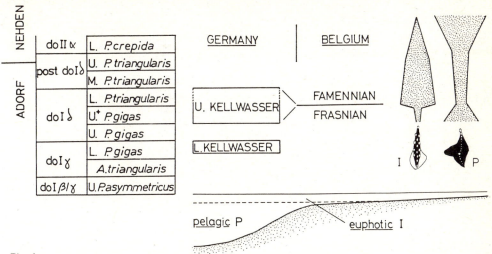

Fig. 4
Conodont biofacies evolution and supposed omission event near the Frasnian-Famennian boundary (I = *Icriodus*, P = *Palmatolepis*). Proposed conodont habitas after Sandberg & Dreesen (1984).

particular black shale facies occurred at the base of a transgression over shallow shelf areas, subsequently to subsiding movements.

The FF-boundary event is evidenced on the Ardennes shelf by a paleontological condensation or even a stratigraphical gap (Fig. 4). It is noteworthy that this local omission event is synchronous with a more important, global event, the Frasnian-Famennian boundary extinction. At, or just before this boundary there is a huge disappearance of biomass and shallow-water benthos over much of the world, and this may have taken place in as short a time as one conodont zone (approx. 0.5 million years).

A short-term low-diversity crisis occurred within the conodont populations as well, at the end of the Uppermost *P. gigas* Zone and was terminated by the end of the Middle *P. triangularis* Zone (Ziegler, 1984). On the neritic shelf (Ardennes), the FF-boundary is characterized by a conspicuous boom of euphotic conodonts (*Icriodus*) and a temporary decrease of pelagic conodonts (*Palmatolepis*), reflecting paleoecologically an eustatic sea-level fluctuation (Dreesen, 1984). Locally, we may even observe an erosional unconformity between the Frasnian and Famennian, as evidenced by the recent observation of an eroded biohermal reef (red "F2"-type mudmound) in the uppermost Frasnian, surmounted by nodular shales of the lowermost Famennian in the former marble quarry at Rance, southern Dinant Synclinorium (Biron et al. 1983).

Oolitic Ironstones

Very intriguing further is the frequent occurrence of Clinton-typ or Minette-type oolitic ironstones and associated marine red beds within nearshore and offshore settings of the Ardennes shelf.

These ironstones occur at distinct stratigraphic levels and represent isochronous marker-beds (dated by conodonts) which can be traced over several tens of kilometers on the shelf (Dreesen, 1982).

A total of 7 dinstinct oolitic ironstone levels are known from the Upper Devonian shelf south and southeast of the Brabant Massif (Fig. 5). With the exception of the lowermost (basal Frasnian) and topmost (Strunian) levels, the oolitic ironstones tend to be concentrated in the Nehden Stufe (Lower and basal Upper Famennian).

During this particular time interval, we observe also an important development of red pelitic sediments in the more pelagic settings of the Rheinisches Schiefergebirge. Moreover, each ironstone level on the shelf seems to correspond biostratigraphically (dated by conodonts and entomozoan ostracods) to distinct levels of turbidites, mud flows, or synsedimentary volcanic deposits (pillow lavas and tuffs) known from different localities in the Rheinisches Schiefergebirge (Fig. 10). The lowermost oolitic ironstone level (0) is isochronous with the exhalative "Roteisenstein Grenzlager" of the Lahn-Dill area (S-border of the Rhein. Schiefergeb.) where it is interstratified within pillow lavas and tuffites ("Grenzdiabas", "Schalstein" and "Tuffe"), whereas the youngest ironstone

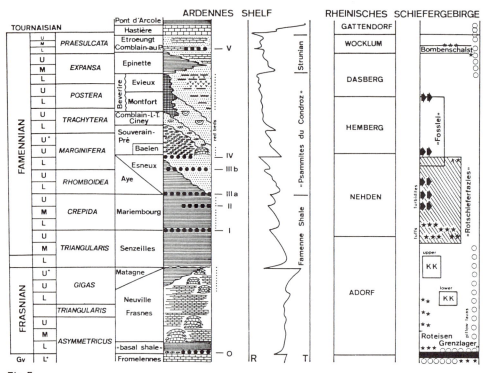

Fig. 5

Even-stratigraphical correlation scheme for the Upper Devonian in the Ardenno-Rhenish Massif. From left to right: Standard conodont zonation amended by Ziegler & Sandberg, 1984 (extreme left) − selected (in) formal lithostratigraphic units of the Ardennes (left column) (legend: see further) − onlap and offlap, relative sealevel fluctuations (middle column) (T: transgression, R: regression) − german stratigraphic units (right column) − episodic events within the "pelagic" settings of the Rheinisches Schiefergebirge (extreme right) (LKK: Lower and UKK; Upper Kellwasser Kalk horizons, black arrows: turbidites, white circles: pillow lavas, black triangles: tuffs, heavy black line: exhalative iron ore). Black dots and roman numerals refer to different oolitic ironstones (after Dreesen, 1982).

level appears to be coeval with a conspicuous volcanic bomb level from the Langenau-bach area (Dreesen & Streel, 1985).

Most frequently the oolitic ironstones occur at the transition of two succeeding litho-stratigraphical units (formations or members). Detailed micropaleontological analysis revealed the presence of paleontological condensation within each ironstone: one cono-dont zone is completely or partly missing, which corresponds to an omission time span of at most 0.5 million years.

Otherwise, field observations and microfacies analysis pointed out that the oolitic iron-stones are allochtonous deposits, resulting from episodic turbulent events.

They consist of heterogenous ferruginized components (ferruginous ooids and related coated grains) which have been derived from (very) shallow nearshore environments and which have been subsequently transported by storm surges or even by more violent processes (tsunamis?) towards the open shelf.

The oolitic ironstones can be petrographically characterized as wackestones or pack-stones with varying amounts of non-sorted, spherical to elongated ferruginous ooids: ferruginous ooid limestone facies, Fe-biomicrites and Fe-biosparites. These ferruginous ooid-bearing wacke/packstones are embedded in nodular shales or micaceous silt- and sandstones. Their thickness varies from more than 100 cm to only a few cm or less, according to their paleogeographic position on the shelf. Most frequently they surmount algal-encrusted and mineralized hardgrounds and/or basal erosional unconformities.

Within a same level the hardgrounds may be repeated. They originated at the end of a re-gression and at the beginning of a new transgression. The lithification itself occurred dur-ing reduced or non-deposition conditions.

The transport mechanism is well-illustrated by the lateral evolution of each ferruginous ooid limestone facies (Fig. 6):

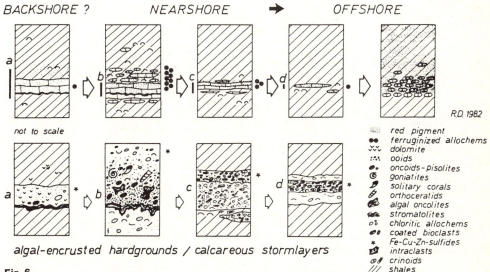

Fig. 6

Idealized scheme for the lateral evolution of a Famennian oolitic ironstone. Upper row: megafacies evolution, lower row: representative microfacies evolution, detail of upper row sections a to d (Dree-sen, 1982).

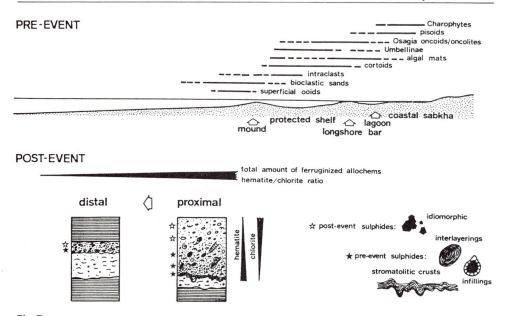

Fig. 7

Ideal distribution of allochems, before their diagenetic replacement by iron minerals and before their removal and subsequent transport. Below: density-stratification of ferruginized allochems, proximal and distal facies, sulphide distribution after transport.

the total amount of ferruginized allochems (superficial ooids, *Osagia*-type oncoids, coated bioclasts and intraclasts, etc.) decreases in an offshore direction. At the same time the thickness of the enveloping limestone is decreasing in the same direction, the total amount of ferruginized coated grains is decreasing from bottom to top, as well as the hematite/chlorite ratio of the ferruginous components (density-stratification or graded bedding). The ferruginous coated grains are heterogenous and they originated in different shallow subtidal, intertidal and even supratidal coastal environments (Fig. 7), from which areas they have been removed by severe storm events. Subsequently they have been redeposited on the shelf, by settling from turbid surficial clouds. The most offshore facies of each oolitic ironstone consists of a few dispersed ferruginized coated grains only or of a red-stainded pelito-carbonate sediment. The red color is due to staining by a fine hematite pigment (minimum 1.5 % of Fe_2O_3).

From the above observations a working model for the origin and distribution of the oolitic ironstones and their coeval red beds can be proposed, comprising four main phases (Fig. 8):

A. In the first phase, calcareous coated grains (superficial ooids, coated skeletal grains,...) accumulated along dispersed, high-energy, fore-barrier shoals, on which algal-sponge-crinoidal carbonate buildups could eventually develop (Dreesen & Flajs, 1984).

B. Small-scale epeirogenic movements resulted in a sea-level fall, non-deposition and even local erosion of the former crinoid accumulations. During this phase hardgrounds formed on the inner shelf, whereas micritization of the above coated grains took place in the peri-shoal area. Otherwise, the more protected "back-mound" shelf area was prolific for the growth of *Osagia*-type oncoids, Umbellinaceans and even Charophycean algae (*Sycidium*).

C. A further withdrawal of the sea (third phase) produced supratidal conditions in coastal sabkhas and evaporitic lagoons, whereas parts of the former shoals became influenced by marine and even meteoric phreatic waters.

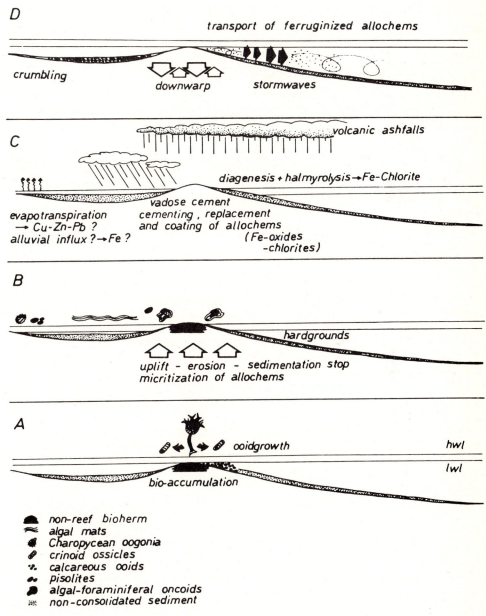

Fig. 8
Sequential stages in the development of Fammenian oolitic ironstones (Dreesen, 1982)

During this phase the carbonate-coated grains became impregnated, coated or replaced by iron minerals: possibly this process took place at the sediment-water interface of the up-lifted near-coastal area, where Fe-sulphides and Fe-chlorites precipitated, grading eventually into (late-diagenetic?) Fe-(hydr)oxides.

The origin of the iron is still a matter of speculation. In contradiction with the "classical" idea of laterite-derived iron (weathering of latosols under humid tropical conditions), the iron in our model is not of lateritic origin. Most probably it required more than one source (Fig. 9): the iron could have been derived from the surface waters of nearby density-stratified evaporitic pans (Sonnenfeld et al., 1977), whereas Fe-sulphides could have replaced some of the allochems within the organic-rich bottom oozes.

The Fe-chlorites could also have originated from the diagenesis and halmyrolysis of volcanic ashes: chloritized bentonite relics, as well as idiomorphic zircons have locally been observed within the enclosing carbonate matrix.

D. Downwarp of the area and subsequent storm-induced transport of the ferruginized coated grains to more offshore settings of the silicoclastic shelf, mark the fourth and final phase.

Within some of the chloritized allochems (preferentially oncoids), relatively high values (up to a few thousand ppm) of base metals (Cu-Zn) have been recorded, whereas polygons of Fe-Cu-Zn sulphides are scattered within the carbonate matrix of the distal ferruginous ooid limestone facies. The former might be related to evaporite-associated metalliferous deposits within the organic-rich lagoonal muds, where base metals became trapped and interstratified within algal mats and oncoids after organic decay, before their removal and transport by storm surges (Figs. 9, 11). The latter sulphides precipitated most probably after the storm-induced transport, and after settling of the finest (clay) fraction from the turbid surficial cloud, to which fraction they were fixed by adsorption.

During the same stratigraphic interval, with maximum development of oolitic ironstones on the Ardennes shelf, red pelites with subordinate sandstone turbidites and volcanic tuffs were deposited in the more pelagic and hemipelagic settings of the Rheinisches Schiefergebirge (Fig. 10). The red pelitic sediments have been interpreted as the result of particular paleogeographic conditions during the sedimentation (Franke & Paul, 1980): decrease of continental detrital influx from the Caledonian source areas, which is compatible with the reduced sedimentation and non-deposition on the shelf (shoaling upward or minor regressive cycles preceeding the formation of each oolitic ironstone).

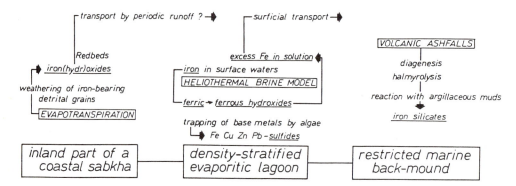

Fig. 9

Possible sources and suggested mechanism for the origin of the iron in the Famennian oolitic ironstones.

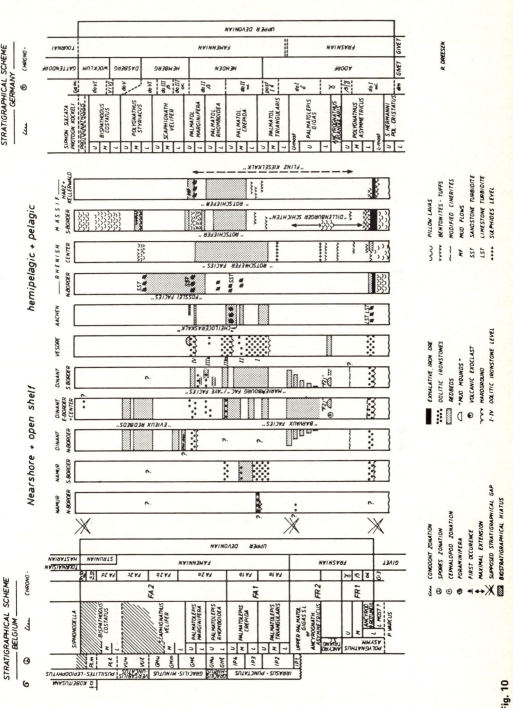

Fig. 10

Correlation scheme for Upper Devonian episodic event deposits within the different tectonic units of the Ardenno-Rhenish Massif (Dreesen, 1982).

The red color is due to the presence of hematite, which has formed during diagenesis from ferric iron bound to the clay fraction. This red pigment could only survive in those pelagic and hemipelagic settings where the organic supply was too low to consume the available oxygen (oligotrophic conditions). This deficit of organic matter during Nehden times is attributed to the paleogeographical conditions mentioned above: the Caledonian relief was greatly reduced by erosion during late Devonian times, and the extent of the source area of the siliciclastics diminished by advancing marine transgressions.

Within these red mudstones, intercalations of sandstone turbidites and of volcanic tuffs are time-equivalent deposits of the oolitic ironstones deposited on the shelf; obviously these turbulent episodic events must have been triggered by a common mechanism.

The turbiditic sandstones are not red because the pigment content has been taken below the critical level (1.5 %) by dilution with non-red siliciclastics.

Interdependence of Episodic Events

Although the Late Devonian generally has been considered as a relatively "quiet" geotectonic period in the "geosynclinal" development of the Ardenno-Rhenish Massif, tectonic movements occurred episodically and have strongly influenced the depositional history of the sedimentary basin south and southeast of the London-Brabant Massif.

Synsedimentary faults and volcanic lineaments anticipating or perpendicular to the tectonic strike directions are apparently the only physical evidences of geotectonic activity in the Rhenohercynian Zone (Walliser, 1981; Franke et al., 1978).

Nevertheless, rapid or short-term sea-level fluctuation (onlap, offlap) and particular episodic event deposits (oolitic ironstones, turbidites, volcanic ashes) are the mute evidences of smaller-scale geotectonic activities during late Devonian times in the Ardenno-Rhenish Massif. All of these events are strongly interdependent and they result from a common pre-orogenic and intra-cratonic tectonic mechanism: epeirogenic (block-faulting) movements of the Paleozoic basement, preferentially along deep-seated faults, which have been reactivated episodically (Lower Devonian through Dinantian) (Fig. 11).

Rifting, block-faulting or tensional movements along deep-seated faults could have easily provoked sea-level fluctuations on a shallow epi-continental shelf, bordering a relatively flat, deeply eroded "continent". A withdrawal of the sea, combined with a reduced influx of siliciclastics allowed the formation of shoaling upward cycles, of condensed sequences and of hardgrounds on the shelf. Moreover, semi-arid climatological conditions on the emerged coastal area would have led to the development of a coastal sabkha complex. Evapotranspiration processes on the inland edge of this coastal sabkha as well as reducing processes in the bottom ooze of density-stratified evaporitic lagoons are potential sources for iron(hydr)oxides and base metal sulphide mineralization. Under these environmental conditions, ferruginous ooids originated most probably through diagenetic replacement of former carbonate coated grains on the different emerging coastal and shallow marine subenvironments.

Highly explosive volcanic activity — as a result of the above tectonic movements — would have produced considerable amounts of volcanic ashes which may have easily reached the Ardennes shelf by eolian transport. After being dropped into the sea these ashes could have been a potential source for chlorites, through chemical reaction with marine argillaceous muds.

Otherwise, explosive volcanism and seismic waves could have triggered diverse turbulent events, such as turbidity currents, severe storms, even tsunamis, which, in their turn, would have been responsible for the transport of both the ferruginized coated grains into the open shelf and the shelf sands into the basinal areas.

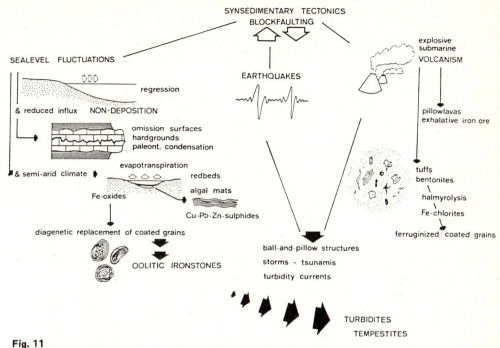

Fig. 11

Cartoon showing the interdependence of tectono-sedimentary processes and the origin of episodic event surfaces and deposits.

Evidence for highly explosive volcanism is found for instance in the occurrence of the so-called "Bombenschalstein" in the basal Wocklum Stufe of the Rhenish Massif (Fig. 5). On the other hand, indirect evidences for seismic activity might be deduced from the episodic formation of ball-and-pillow structures ("pseudo-nodules") within the longshore barrier sands complex of the Montfort Formation (Hemberg Stufe) in the Ardennes. The origin of such deformation structures has recently been related to liquefaction of underlying clayey silt and its subsequent loss of load-bearing capacity in response to the passage of a seismic wave (Hempton & Dewey, 1983).

Oolitic ironstones and coeval sandstone turbidites or volcanic tuffs thus represent excellent event-stratigraphical marker beds. They are the mute evidences of episodic events, which might be related to pre-orogenic, intra-cratonic synsedimentary tectonics within the Ardenno-Rhenish Massif.

Acknowledgements

The manuscript was written at the Geological Institute of the RWTH Aachen, where Prof. W. Kasig provided working facilities. The work was done during the tenure of a research position granted by the Alexander von Humboldt Foundation. The manuscript benefited from comments by Prof. J. Thorez (Liège).

References

Biron, J.P., Coen-Aubert, M., Dreesen, R., Ducarme, B., Groessens, E., & Tourneur, F. (1983): Le trou de Versailles ou Carrière à Roc de Rance. Bull. Soc. belge de Géol., 92, 4, 317—336.

Copper, P. (1984): Cold water oceans and the Frasnian-Famennian extinction crisis. Geol. Soc. America Abstracts with Programs. Abstr. nr. 43385.

Dott, R.H. (1982): Episodic view now replacing catastrophism. Geotimes, 27, 11, Nov. 1982, 16—17.

Dreesen, R. (1982): Storm-generated oolitic ironstones in the Famennian (Falb-Fa2a) of the Vesdre and Dinant Basins (Upper Devonian, Belgium). Ann. Soc. géol. Belg., 105, 105—130.

Dreesen, R. (1984): Stratigraphic correlation of oolitic ironstones in the Verviers and Havelange Bore-holes (Upper Devonian, Vesdre and Dinant Synclinoria, Belgium). Bull. Soc. belge de Géol., 93, 1—2, 197—211.

Dreesen, R. & Flajs, G. (1984): The "Marbre rouge de Baelen", an important algal-sponge-crinoidal carbonate buildup in the Upper Devonian of the Vesdre Massif (E-Belgium). C.R. Acad. Sci. Paris, 299, série II, 10, 639—644.

Dreesen, R. & Thorez, J. (1980): Sedimentary environments, conodont biofacies and paleoecology of the Belgian Famennian — an approach. Ann. Soc. géol. Belg., 103, 97—110.

Dreesen, R. & Thorez, J. (1982): Upper Devonian sediments in the Ardenno-Rhenish area: sedimentology and geochemistry, in: "The Pre-Permian around the Brabant Massif", Abstracts of the Third Int. Coll. Maastricht April 1982, Public. Natuurhist. Gen. Limburg, 1982, 32, 1—4, 8—15.

Dreesen, R. & Streel, M. (1985): A new event-stratigraphical marker bed in the Uppermost Devonian of the Ardenno-Rhenish Massif (abstract). Ann. Soc. géol. Belg., 108, 412.

Dvorak, J. (1973): Die Quer-Gliederung des Rheinischen Schiefergebirges und die Tektogenese des Siegener Antiklinoriums. N. Jb. Geol. Paläont. Abh., 143, 2, 133—152.

Einsele, G. & Seilacher, A. (1982): (Editors) Cyclic and Event Stratification, Springer Verlag, Berlin. Heidelberg, New York, 536 p.

Franke, W, Eder, W., Engel, W. & Langenstrassen, F. (1978): Main aspects of geosynclinal sedimentation in the Rhenohercynian Zone. Z. dt. geol. Ges., 129, 201—216.

Franke, W. & Paul J. (1980): Pelagic redbeds in the Devonian of Germany. Deposition and diagenesis. Sed. Geol., 25, 231—256.

Hempton, H.R. & Dewey, J.F. (1983): Earthquake-induced deformational structures in young lacustrine sediments, East Anatolian Fault, Southeast Turkey. Tectonophysics, 98 (1983), T7—T14.

Krebs, W. (1969): Über Schwarzschiefer und bituminöse Kalke im miteleuropäischen Variscikum. Erdöl u. Kohle, 22, 2—6, 62—67.

Krebs, W. (1979): Devonian basinal facies, in: "The Devonian System". Spec. Papers in Palaeontology, 23, 125—139.

McLaren, D.J. (1982): Frasnian-Famennian extinctions. Geol. Soc. America Special Paper 190, 477—484.

McLaren, D.J. (1983): Bolides and Biostratigraphy. Geol. Soc. America Bull. 94, 313—324.

McLaren, D.J. (1984): An Upper Devonian Event: Frasnian-Famennian extinctions. Geol. Soc. America Abstracts with Programs, nr. 42323.

Sandberg, C. A. & Dreesen, R. (1984): Late Devonian icriodontid biofacies models and alternate shallow-water conodont zonation, in: Clark, D. L., ed., Conodont biofacies and Provincialism. Geological Society of America Special Paper 196, 143—178.

Sonnenfeld, P., Hudec, P.P. & Turek, A. (1977): Base metal concentration in a density-stratified evaporitic pan. A.A.P.G. Studies in Geology 5, 181—187.

Thorez, J., Streel, M., Bouckaert, J. & Bless, M.J.M. (1977): Stratigraphie et paléogéographie de la partie orientale du Synclinorium de Dinant (Belgique) au Famennien Supérieur: un modèle de bassin sédimentaire reconstitueé par analyse pluridisciplinaire sédimentologique et micropaléontologique. Meded. Rijks Geol. Dienst. Ns, 28, 2, 17—28.

Walliser, O. (1981): The geosynclinal development of the Rheinisches Schiefergebirge (Rhenohercynian Zone of the Variscides, Germany). Geol. Mijnbouw, 60, 89—96.

Ziegler, W. (1984): Conodonts at the Frasnian/Famennian crisis. Geol. Soc. America Abstracts with Programs, nr. 43387.

Ziegler, W. & Sandberg, C. A. (1984): Palmatolepis-based revision of upper part of standard Late Devonian conodont zonation, in: Clark, D. L., ed., Conodont biofacies and Provincialism. Geological Society of America Special Paper 196, 179—194.

Heavy Mineral Analysis on Lower Devonian Rocks of the Ebbe-Anticline (Rheinisches Schiefergebirge)

W.P. Loske/H. Miller

Institut für Allgemeine und Angewandte Geologie der Universität, Luisenstr. 37, D-8000 München 2, FRG

Key Words

Heavy mineral analysis
Ebbe-Anticline
Sedimentation model

Abstract

The statistical analysis of heavy minerals on rock samples from the Lower Devonian of the Ebbe anticline displays a correlation between the quantity of heavy minerals found in the analysed rocks and the estimated subsidence and sedimentation rates. There is also a difference in the quantitiy of heavy minerals in rocks formed under marine (low rate) or terrestrial (high rate) conditions.

A sedimentation model for the Ebbe-Anticline during Early Devonian times can be developed from a zircon analysis. Three types of zircon populations can be recognized: polycyclic-granitic, sedimentary-metamorphic, and volcanic. The first two populations represent two different types of allochthonous source regions having fed the sedimentation area. Their variation in frequency is interpreted as a result of geological changes during sedimentation (e. g. sediment distribution processes, sediment drift, marine-terrestrial conditions). The volcanic population derived from the local keratophyre volcanism. Therefore it only locally contributes to the sedimentary processes.

Introduction

During the past years only a few papers have been published dealing with heavy minerals of the Sauerland. Some used the heavy mineral analysis to bring light into stratigraphic problems, others to solve the question of the origin of the sediments in the Rhenish Slate Mountains (e. g. Degens 1955; Henningsen 1961, 1963; Homrighausen 1979).

In sedimentary rocks in which the heavy mineral spectrum consists only of some few stable minerals like zircon, tourmaline or rutile, a mineral analysis based on morphologic characteristics can be very useful to get more information. For zircons this method has recently been used by Pupin et al. (1969), Zimmerle (1979), Trautnitz (1980), Brix (1981) and Winter (1981).

The Ebbe Anticline is a WSW-ENE striking structure between Wipperfürth in the west and Plettenberg in the east (Fig. 1). For our purpose we selected a suite of mostly fine grained, arenaceous rocks from the Bredeneck Formation up to the Remscheid Formation. The petrography of the rocks has been described by Fuchs (1928), Fuchs & Schmidt, W.E. (1928), Ziegler (1970, 1978), Böger (1981, 1983), Timm (1981) and Timm, Degens & Wiesner (1981).

The Technique of Heavy Mineral and Zircon Analysis

The preparation of the collected samples shall be explained only briefly; a full description is given in Loske (1983). It consists in:
— Cleaning and drying of the samples;
— Cracking in a jaw breaker;
— Squeezing in a roll mill;
— Fractionating into a grain size range of 0.16—0.04 mm;
— Removal of iron hydroxide crusts;
— Separation of the heavy minerals in a centrifuge with bromoform as sedimentation liquid;
— Mounting of the separated heavy minerals on an object carrier with an artificial resin (piperine n = 1.68).
For the heavy mineral analyses, approximately 200—300 non-opaque grains were counted in each sample. Besides zircon, tourmaline, rutile and picotite we found: anatase, apatite, brookite, epidote, garnet, monazite, sphene, xenotime, sphalerite and zoisite. But only the first four were found in sufficient quantity in all samples. For statistical analyses the counted frequencies of the minerals were transformed to a value that describes the quantity (mass) of the heavy minerals in each sample standardized on one ton of rock.
The zircon analysis was carried out on 25 different characteristics grouped as follows:
1. Shape
— roundness
— form (Tracht)
— growth hindrances
— scarred surfaces
— sheety overgrowth
— single-sided newgrowth
2. Internal structures
— cores
— zoning
— bubbly inclusions (vesiculous)
— solid inclusions
— cracks
— cracking of the core
3. Color
— dim
— partly dim
— reddish color
— brownish color
In each sample we registered the characteristics of 100 zircon grains. Then we computed the frequency of each single characteristic (25) and of 11 combinations of certain characteristics, which are significant for the identification of the former host rock of the zircons. Table 1 presents the collection of all zircon characteristics used in the statistical analysis.

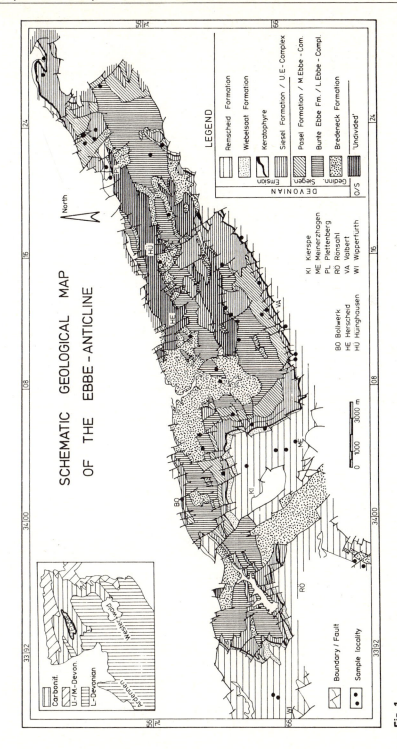

Fig. 1

Schematic geological map of the Ebbe-Anticline. The sandstone outcrops around r-3400/h-7466 after new concepts (Börding, pers. commun.) are of Siegenian and Early Emsian age.

Table 1: Tabulation of all zircon characteristics used in the statistical analysis (25 simple zircon characteristics and 11 combined zircon characteristics)

No. of the variables used for clustery-anlaysis	No. of the character-istic or characterist-ics-combination	Short description of the zircon characteristics
1	1	reddish color
2	2	dim
3	3	partly dim
4	4	brownish color
5	5	colorless core
6	6	colored core
7	7	indirectly visible core
8	8	sheety overgrowth
9	9	single-sided newgrowth
10	10	well rounded
11	11	medium rounded
12	12	badly rounded
13	13	idiomorphic
14	14	crystal edges
15	15	zoning
16	16	cracking of the core
17	17	cracks
18	18	bubbly inclusions (vesiculous)
19	19	solid inclusions
20	20	growth hindrances
21	21	form 1
22	22	form 2
23	23	form 3
24	24	scarred surface
25	25	pimply newgrowth
26	1—10	reddish/well rounded
27	1—11	reddish/medium rounded
28	9—11	newgrowth/medium rounded
29	11—14	med. rounded/crystal edges
30	12—14	badly rounded/crystal edges
31	12—18	badly rounded/bubbly inclusions
32	12—19	badly rounded/solid inclusions
33	13—14	idiomorphic/crystal edges
34	15—18	zoning/bubbly inclusions
35	15—19	zoning/solid inclusions
36	18—19	bubbly/solid inclusions

Geological Interpretation of the Factor and Cluster Analyses

Heavy Minerals

The factor analysis was carried out with the aid of the SPSS-software package of the com-
puter center of the University of Münster (subroutine FACTOR Type PAL; Nie et al.
1975). The factor analysis with heavy mineral data led to the following results:
The positive loading of the four measured heavy minerals on the first factor displays that
all samples have in common as a positive community their contents of heavy minerals related to a ton of rock. In a geological sense this means heavy mineral concentration (accumulation) or dilution processes. With respect to the local processes in the sedimentation
area this factor can be described by terms like "supply of detritus", "subsidence rates"
and "facies". That is why we named it "environment factor". The second factor displays

that tourmaline and picotite are inversely correlated, whereas zircon and rutile show no dependence on this factor. Obviously this factor depends on the lithology of the area supplying the detritus. During times of higher supply of picotite bearing rocks, the relative proportion of tourmaline decreases and vice versa. In a more general way this phenomenon can be paraphrased as a "composition of detritus-type dependence".

For a cluster analysis we used the software package "CAM" (Clusteranalyse Münster) described by De Lange & Steinhausen (1981).

For the interpretation of the cluster analysis, the area was divided into three regions: a westerly one, called Meinerzhagen; a central part, named Herscheid; and an easterly one, called Plettenberg. The samples were grouped into three clusters. Cluster 1 contains all samples with a low heavy-mineral rate and clusters 2 and 3 all samples with a medium to high heavy-mineral rate. A trendplot (Fig. 2) gives an idea of the distribution of the samples in a stratigraphical and regional view. Compared with the thickness of all stratigraphic units (taken from Oncken 1982), a negative correlation is recognizable (Fig. 3–6). This means if the thickness of a unit is high, the heavy mineral contents of the rocks is quite low and vice versa. As a second factor the conditions during sedimentation influenced the heavy mineral contents. There is an evident difference between the high heavy-mineral frequency in the terrestrially formed Siesel Formation (Fig. 5) and the rather low heavy-mineral rate in the mostly marine Remscheid Formation (Fig. 6).

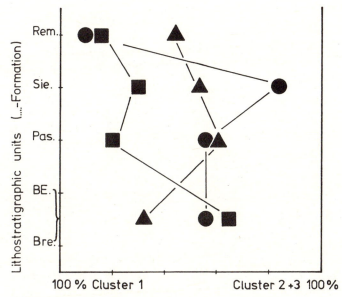

Fig. 2

Trendplot of the heavy mineral distribution
western region: Meinerzhagen (squares)
central region: Herscheid (circles)
eastern region: Plettenberg (triangles)
cluster 1: low heavy-mineral rate
cluster 2 + 3: medium to high heavy-mineral rate

Fig. 3—6

Thickness and heavy mineral rate (The isopachs and tectonic elements are taken from Oncken 1982). The density of the dotting is equivalent to the heavy mineral rate

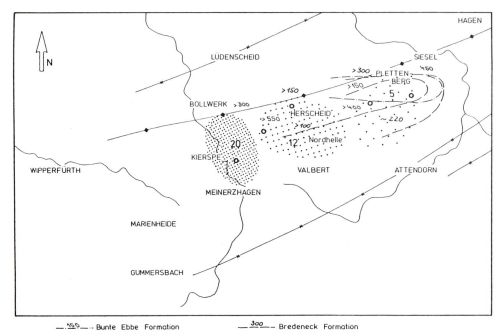

—.ⁱ⁵⁰—.- Bunte Ebbe Formation   —300— Bredeneck Formation

Fig. 3

Thickness and heavy mineral rate in the Bredeneck Formation and Bunte Ebbe Formation

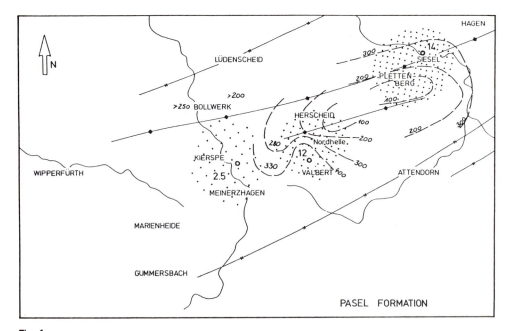

PASEL FORMATION

Fig. 4

Thickness and heavy mineral rate in the Pasel Formation

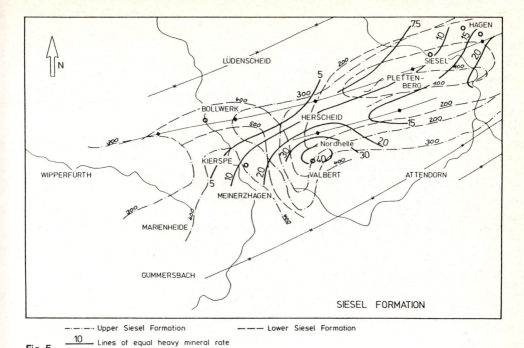

Fig. 5

Thickness and heavy mineral rate in the upper and lower Siesel Formation
unbroken lines: lines of equal heavy mineral rate
broken lines: isopachs

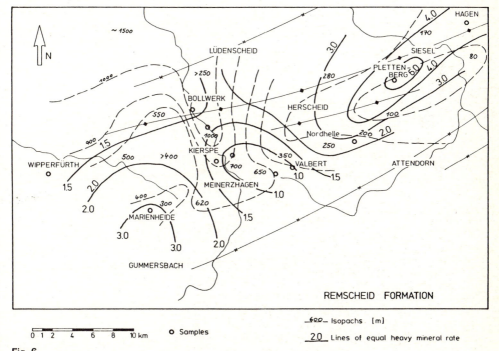

Fig. 6

Thickness and heavy mineral rate in the Remscheid Formation (see Fig. 5)

Zircon Characteristics

A factor analysis was made from the zircon characteristics described in Table 1.

The first factor supports all characteristics of a fresh, volcanic zircon population with a high positive loading. A polycyclic-granitic population is described by a second factor, and a third factor describes the characteristics of zircons of metamorphic origin. Two further factors describe another granitic population and a population with sedimentary characteristics. But we emphasize that these last two factors are not very trustworthy, because their eigenvalues are less than five percent.

The cluster analysis of the data shows the same five zircon populations: a polycyclic-granitic population divided into two clusters, a volcanic population also divided into two clusters, and a single cluster with zircons of metamorphic origin.

The trendplots (Fig. 7—9) display that the samples with a predominance of polycyclic-granitic zircons are mostly found in the regions of Herscheid and Plettenberg (Fig. 7). The samples with a predominance of metamorphic zircons (Fig. 8) occur in the regions of Herscheid and Meinerzhagen. All three regions have in common a relatively high percentage of a volcanic component during the Emsian (Fig. 9). We believe that the volcanic debris derived from the local keratophyre volcanism. Therefore it is necessary to eliminate this debris when trying to get information about the development of the allochthonous detritus sources. This is done in Fig. 10.

Fig. 7—9

Trendplots of the zircon cluster analysis (relative percentage of the samples of each region and stratigraphic unit; the symbols are the same as in Fig. 2)

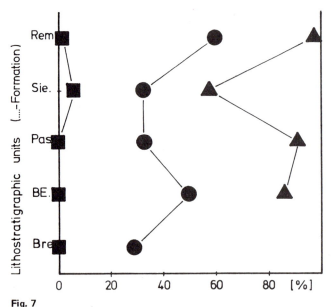

Fig. 7

Trendplot of cluster 1 + 5 (polycyclic-granitic zircon population)

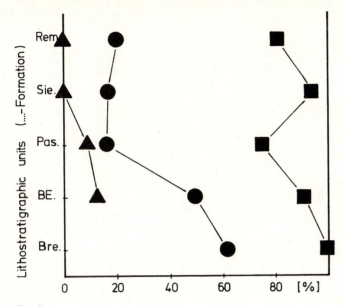

Fig. 8

Trendplot of cluster 2 (metamorphic zircon population)

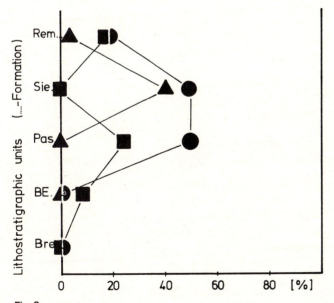

Fig. 9

Trendplot of cluster 3 + 4 (volcanic zircon population)

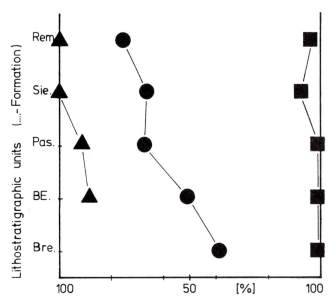

Fig. 10

Trendplot of the allochthonous zircon populations
left: eastern source region (= cluster 1 + 5)
right: western source region (=cluster 2)
Obviously the Meinerzhagen region (squares) is fed by the debris
from a western source region (right), whilst the region Plettenberg
(triangles) is fed by the debris of the eastern source (left). The region Herscheid (circles) is influenced by both sources.

The model of sedimentation discussed now (Fig. 11–13) is based on two rivers coming
from northern directions. Their sedimentary discharge is distributed in broad delta fans
(Böger 1983). The western one gets its debris from a metamorphic source region,
whereas the easterly one is fed by a more granitic, polycyclic zircons-bearing lithologic
unit.

In the Bredeneck Formation (Fig. 11), the distribution of the debris is controlled by
marine currents. Böger (1981) describes the upper Bredeneck Formation as built on
the proximal part of a delta front. The currents must have initiated a drift into easterly
directions, because the trendplot (Fig. 10) displays a significant influence of the debris
from the west onto the eastern region of Plettenberg. During the formation of the Bunte
Ebbe Formation, the terrestrial history of the Ebbe Anticline began. These strata were
formed in backwash areas on lower courses of rivers (Timm 1981). But the sedimentation must still have taken place under aquatic conditions, because subaerial sedimentation indicating marks like drying cracks are rather rare (Timm 1981). The debris drifted
into eastern directions as well, because the western zircon association is found in the eastern
region of Plettenberg, but not vice versa. This state of evolution of the Ebbe area was still
present during the sedimentation of the Pasel Formation, which was formed on a flood
plain (Böger 1981). But during the sedimentation of the Siesel Formation, a characteristic

change is obvious. Not only rocks like conglomerates or red shales are numerous now, but also the direction of sediment drift must have changed (Fig. 12). The influence of the debris from the western source region disappears in the eastern region. Now in the western region of Meinerzhagen, a lot of samples seem to be influenced by the eastern debris source (Fig. 10).

After the K4 volcanic event the terrestrial history of the Ebbe area closed down, and after a transition period (documented by black and colored shales) marine environments returned. Marine fossils, bioturbations and frequent flaser bedding indicate a shallow sea of intertidal conditions. As can be seen in Fig. 10, the debris of the eastern source is no longer visible in the region of Meinerzhagen, but is still dominating in the region of Herscheid. This indicates again a westerly drift of the debris (Fig. 13). The rather small influence of the debris from the western source in the region of Herscheid is easily explained by the existence of the Meinerzhagen trench (sensu Oncken 1982).

Fig. 11—13

Schematic sedimentation model for the Ebbe region

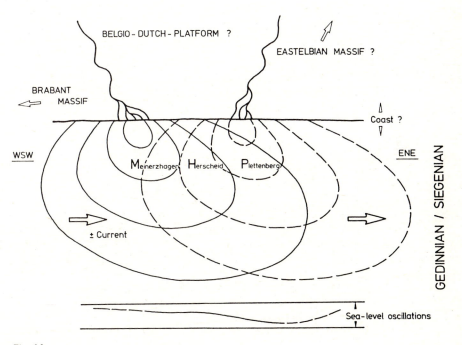

Fig. 11

Bredeneck Formation up to Pasel Formation

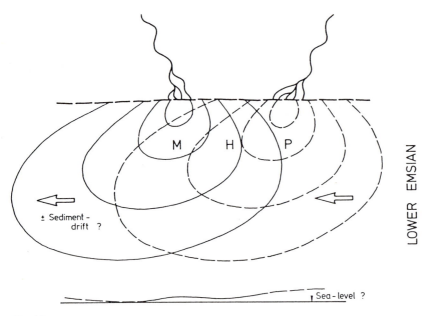

Fig. 12
Siesel Formation (terrestrial environment)

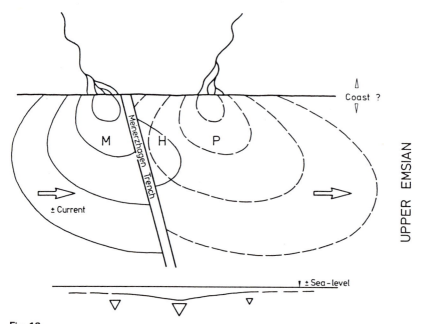

Fig. 13
Remscheid Formation (marine environment)

Conclusions

A heavy mineral and zircon analysis was carried out on samples from the Lower Devonian of the Ebbe Anticline. Except for zircon, tourmaline, rutile and picotite, all other heavy minerals found in the samples are below a significant level. A factor analysis, made with data computed from the frequency and density of the four stable minerals, displays that the first factor can be described as an "environment factor" giving information about the quantity of heavy minerals in the analysed rocks. The second factor depends on the quantitative ratios of the different heavy minerals. This factor is called "composition" or "discharge-type factor".

A Q-mode cluster analysis points to the fact that, except for local accumulation processes, the subsidence-rate and the rate of sedimentation are inversely proportional to the quantity of the non-opaque heavy minerals in the sediment. A fundamental difference is shown in the quantity of heavy minerals from rocks of marine origin and from rocks of terrestrial origin (Ebbe-Komplex). The significant appearance of picotite in some regions from the Ebbe Anticline points to a contribution of a basic or ultrabasic motherrock.

A zircon analysis is based on 25 different characteristics of grains, describing their morphology and their internal structures. The factor analysis displays that although the characteristics of the individually described zircons were reduced to the general frequency of the characteristics in a sample, there is still the possibility of differentiating several types of zircon populations. In general one zircon population derived from volcanic origin, another one of granitic polycyclic origin, and a third one of metamorphic origin can be recognized.

A comparable group of zircons of volcanic, plutonic and metamorphic origin results from a cluster analysis. The stratigraphical analysis of the different clusters in a three-region-model leads to trendplots displaying a significant variation of the different zircon populations in each region. So we are able to evolve a sedimentation model considering the fact that a polycyclic-granitic population and a sedimentary-metamorphic zircon population represent two different debris types feeding the sedimentation area during Early Devonian times. A volcanic population contributes only to local phenomena, because it is derived from the autochthonous keratophyre volcanism of the Ebbe area. The polycyclic-granitic one can be derived from the Baltic Shield in the north of the basin, where root-zones of Precambrian orogenes are cropping out. The sedimentary-metamorphic one is supposed to come from the Caledonian orogenic belt in the northwest of the basin.

Acknowledgements

This study was funded by the Ministerium für Wissenschaft und Forschung (Ministry for Science and Research) of Nordrhein-Westfalen.

References

Böger, H. (1981): Stratigraphische, fazielle und tektonische Zusammenhänge im Unterdevon des Sauerlandes (Rheinisches Schiefergebirge) und der Kaledonisch-Variszische Umschwung. Mitt. geol.-paläont. Inst. Univ. Hamb. 50, 45−58.

− (1983): Eine Lithostratigraphie des Unterdevons im Sauerlande und im östlichen Bergischen Lande (Rheinisches Schiefergebirge) II. Das Ebbe-Antiklinorium. N. Jb. Geol. Paläont. Abh. 166 (2), 294−326.

Brix, M. (1981): Schwermineralanalyse und andere sedimentologische Untersuchungen als Beitrag zur Rekonstruktion der strukturellen Entwicklung des westlichen Hohen Atlas/Marokko. Diss. Univ. Bonn 247 p.

Degens, E. (1955): Stratigraphie, Tektonik und hydrothermale Vererzung im Raume Wissen-Morsbach. Geol. Rdsch. 44, 391−421.

Fuchs, A., (1928): Geologische Karte von Preußen und benachbarten Bundesstaaten, Blatt Wipperfürth, Maßstab 1:25 000, mit Erläuterungen. 64 S.

Fuchs, A. and W.E. Schmidt, (1928): Geologische Karte von Preußen und benachbarten Bundesstaaten, Blatt Gummersbach, Maßstab 1:25000, mit Erläuterungen.

Henningsen, D., (1961): Untersuchungen über Stoffbestand und Paläogeographie der Gießener Grauwacke. Geol. Rdsch. 51, 600−626.

— (1963): Zur Herkunft und Unterscheidung der sandigen Gesteine am Südostrand des Rheinischen Schiefergebirges. N. Jb. Geol. Paläont. Mh. 1963, 49−67.

Homrighausen, R., (1979): Petrographische Untersuchungen an sandigen Gesteinen der Hörre-Zone (Rheinisches Schiefergebirge, Oberdevon-Unterkarbon). Geol. Abh. Hess. 79, 1−84.

Lange, N. de and D. Steinhausen, (1981): Programme zur automatischen Klassifikation − Verfahren zur Clusterung quantitativer und qualitativer Daten. Schriftenreihe des Rechenzentrums der Univ. Münster 36, 52 p.

Loske, W.P., (1983): Schwermineral- und Zirkonvarietätenanalyse als Beitrag zur Rekonstruktion der Paläogeographie im Unterdevon des Ebbe-Sattels (Rheinisches Schiefergebirge). Diss. Univ. Münster 198 p.

Nie, N.H., C.H. Hull, J.G. Jenkins, K. Steinbrenner and D.H. Bent, (1975): SPSS-Statistical packages for the social sciences. 675 p., New York.

Oncken, O., (1982): Determinierung und Entwicklung großtektonischer Strukturen im nördlichen Rhenoherzynikum (Beispiel Ebbe-Antiklinorium). Diss. Univ. Köln 189 p.

Pupin, J.P., M. Boucarut, G. Turco and S. Gueirard, (1969): Les zircons des granites et migmatites du Massif de l'Argentera-Mercantour et leur signification petrogenetique. Bull. Soc. France Mineral. Cristallogr. 92, 472−483.

Timm, J., (1981): Die Faziesentwicklung der ältesten Schichten des Ebbe-Antiklinoriums. Mitt. geol.-paläont. Inst. Univ. Hamb. 50, 147−173.

Timm, J., E.T. Degens and M.G. Wiesner, (1981): Erläuterungen zur Geologischen Karte des zentralen Ebbe-Antiklinoriums 1: 25000. Mitt. geol.-paläont. Inst. Univ. Hamb. 50, 59−75.

Trautnitz, H.-M., (1980): Zirkonstratigraphie nach vergleichender morphologischer Analyse und statistischen Rechenverfahren − dargestellt am Beispiel klastischer Gesteine im Harz. Diss. Univ. Erlangen 159 p.

Winter, J., (1981): Exakte tephrostratigraphische Korrelation mit morphologisch differenzierten Zirkonpopulationen (Grenzbereich Unter-/Mitteldevon, Eifel-Ardenen). N. Jb. Geol. Paläont. Abh. 162, 97−136.

Ziegler, W., (1970): Geologische Karte von NRW 1:25 000; Erläuterungen zu Blatt 4713 Plettenberg. Geol. Landesamt NRW 179 p., Krefeld.

— (1978): Erläuterungen zur geologischen Karte von NRW 1:25 000, Blatt 4813 Attendorn. Geol. Landesamt NRW 230 p., Krefeld.

Zimmerle, W., (1979): Accessory zircon from a rhyolithe, Yellowstone National Park (Wyoming, U.S.A). Z. dtsch. geol. Ges. 130, 361−369.

Meta Alkali Basaltic Volcanics do Occur in the Palaeozoic of the Rhenish Mountains

H. D. Nesbor / H. Flick

Geologisch-Paläontologisches Institut der Universität Heidelberg, im Neuenheimer Feld 234, 6900 Heidelberg 1, Federal Republic of Germany

Key Words

Rhenish Mountains
Lahn syncline
Palaeozoic mafic volcanism
Spilitic dolerite sill
Meta alkali basalt
Geotectonic position

Abstract

A spilitic dolerite sill is described from Wasenbach valley at the southwestern end of the Lahn syncline in the southern Rhenish Mountains. The emplacement of the sill was guided by a smaller previously intruded sill which forms together with its contact zone the roof of the larger intrusion. Differentiation during crystallization was rather weak with the exception of a final fraction of latitic composition. The sill experienced secondary alterations (mainly spilitization) and slight metamorphism during Hercynian orogeny. The mineralogical composition leads to the conclusion of an alkali basaltic origin. The other mafic Palaeozoic volcanics of the Lahn and Dill synclines are generally being interpreted as intraplate tholeiites to ocean ridge tholeiites solely on geochemical grounds. The ocean ridge tholeiites should not be regarded as part of the Lahn-Dill complex because they occur in a separate tectonostratigraphic unit (Gießen greywacke nappe). The finding of alkali basaltic rocks is another indication that the Lahn-Dill volcanics are solely of intraplate origin.

Introduction

Palaeozoic sedimentation was accompanied by widerspread volcanism in the Lahn and Dill synclines of the southern Rhenish Mountains. The basic volcanics outnumber the siliceous ones significantly. There were three phases of basic volcanism, two major ones during Givetian to Adorfian time (late Middle Devonian to early Late Devonian) and Early Carboniferous, and a minor one during higher Late Devonian. Products of the basic volcanism include pyroclastics as well as submarine flows, partly with pillows, and sub-

volcanic intrusives. Later alterations have led to spilitization of the volcanics and pala-
gonitic tuffs, the latter being called "Schalstein".

The most recent thorough petrographic description on these volcanics has been given by
Hentschel (1970). Their geochemistry has been investigated and reviewed by Wedepohl
et al. (1983). According to the last account the basic volcanics have intraplate tholeiitic
and ocean ridge tholeiitic character. This seems to be too generalized because alkaline
tendencies are clearly present.

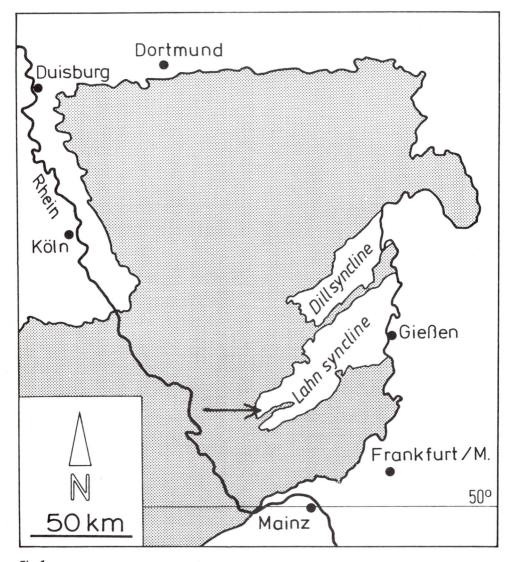

Fig. 1
Map of the Rhenish Mountains with Lahn and Dill synclines. Position of Wasenbach dolerite marked
by arrow.

Petrographic investigations of a doleritic spilitized sill at Wasenbach near Diez/Lahn at the southwestern end of the Lahn syncline demonstrate an alkali basaltic origin of this occurrence. Thus the range of composition of the volcanic rocks is larger than generally accepted. A detailed description of the occurrence will be published elsewhere.

Geological Setting

The spilitic sill of Wasenbach is situated at the southwest end of the large Lahn syncline (Fig. 1) about 10 km southwest of Diez. Specifically, the sill is part of the minor Schaumburg syncline (in the sense of Ahlburg as presented by Kegel 1922). The sill can be followed over a distance of 1,2 km from Rupbach valley in the southwest to the northeast at Steinsberg Küppel (Fig. 2). It has a maximum width of about 65 m near its northeastern end where it crosses the valley of Wasenbach. At this point the sill is very well exposed in two working quarries on both sides of the valley.

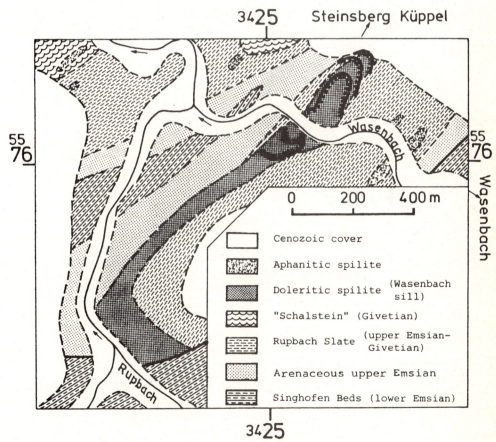

3425 Steinsberg Küppel

55
76

55
76

Wasenbach

Wasenbach

Rupbach

0   200   400 m

☐ Cenozoic cover

▨ Aphanitic spilite

◾ Doleritic spilite (Wasenbach sill)

〰 "Schalstein" (Givetian)

▤ Rupbach Slate (upper Emsian-Givetian)

▦ Arenaceous upper Emsian

▤ Singhofen Beds (lower Emsian)

3425

Fig. 2
Geological map of the area around Wasenbach dolerite sill.

The Wasenbach sill intruded between arenaceous sediments at the bottom and argillaceous sediments at the top. Both lithological units belong to the very Late Lower Devonian (Requadt & Weddige 1978). An exception exists at the northeastern end where the sill lies totally within the argillaceous sediments (Fig. 2). This facies of the Wissenbach slates is called Rupbach slates in this region (Requardt & Weddige 1978). In this particular site the sill coincides with the stratigraphic boundary between Early and Middle Devonian.

Besides the doleritic Wasenbach sill, there exist several smaller aphanitic sills within the surrounding sediments (mainly Rupbach slates). These smaller sills are generally not more than a few meters thick and show a quite different petrography. A few tuffaceous horizons of felsic composition can be found especially in the hanging wall of Wasenbach sill.

Description of Wasenbach sill

Wasenbach sill dips generally about 50° southeast because of its structural position. Jointing is associated partly with the tectonic setting, partly with the intrusion. In the quarry southwest of Wasenbach valley, very coarse columnar jointing can be recognized, the columns being several meters in diameter.

Contact metamorphism on both sides prove the intrusive nature of Wasenbach sill. The argillaceous sediments have been hardened and roughened due to coarsening of the grains together with formation of new crystals of mica and garnet.

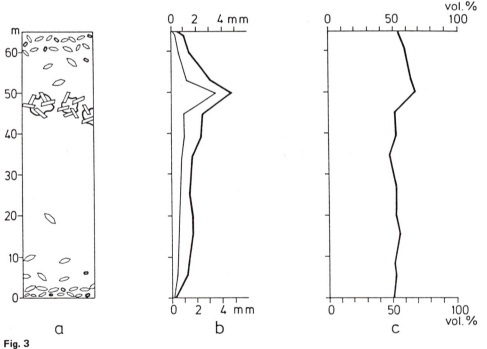

Fig. 3

Development of Wasenbach sill from bottom to top with (a) texture, (b) grain size of plagicolase (thick) and pyroxene (thin), and (c) the portion of salic components. The texture is intergranular throughout (white), porphyritic near the contact (pyroxene with thick outlines, plagioclase with thin outlines) and a tendency to radiate intergrowth at the end of crystallization between 40 and 50m.

Wasenbach sill is formed by a tough, homogenous rock of blackish to greyish green color with doleritic texture. It contains quite a few dikelets of a light grey color. Every now and then small lumps of totally recrystallized country rock up to 20 cm in diameter can be found. Originally, these must have been calcareous shales. The argillaceous parts were totally altered to alkali feldspar, alkali amphibole and alkali pyroxene, the calcareous lenses being macroscopically unaffected.

The grain size of the minerals in the rock shows a clear dependence on the position with a gradual increase from the contact to the upper third of the sill (Fig. 3b). Within the upper third there is a very strong increase in grain size. The distribution of the different mineral phases shows an analogue behavior (Fig. 3c + 4). Similarly, there is a difference in the development of the texture (Fig. 3a) being mainly intergranular. Within the upper third there is a tendency to radiate intergrowth of plagioclase and pyroxene. On the other hand, near the borders of the sill phenocrysts of plagioclase and pyroxene can be found. These increase toward the contact, giving rise to a porphyritic texture.

Mineralogical Description

Plagioclase, clinopyroxene, chlorite, and titanomagnetite are the main constituents within the Wasenbach dolerite. Minor constituents and accessory minerals occur in rather variable amounts. In decreasing sequence they are albite, apatite, amphibole, epidote, stilpnomelane, aegirine, biotite, calcite, and sulfides. Fissures are filled with calcite, chlorite, and partly albite or occasionally with asbestos and albite.

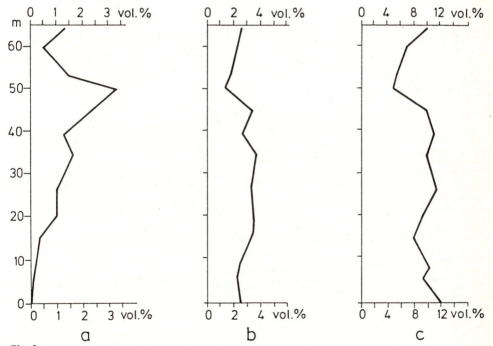

Fig. 4
Portion of (a) amphibole, (b) apatite, and (c) titanomagnetite within Wasenbach sill.

From all the minerals *plagioclase* is the most important one with 50 to 70 vol-% ranging between 0.5 and 8 mm in length (Fig. 3b). The mostly hypidiomorphic crystals are rather clouded and twinned after the albite and pericline law, more rarely after the Karlsbad law. Replacement by chlorite, epidote, occasionally by calcite occurred mainly in the centers of the crystals. The An content reaches only 5 mol-%, the optic angle 2 Vz ranging between 84° and 95°. The plagioclase grains always have low temperature optics, with frequent occurrence of chessboard albite (Fig. 5).

Occasionally, the plagioclase is rimmed by clear *albite* which fills interstitial spaces as well. Idiomorphic albite can be found within joints as irregular intergrown laths.

The *pyroxenes* take up 9 to 20 vol-% in the rock and are of the monoclinic sub-group. They belong dominantly to the augites. Mostly, the crystals show euhedral outlines and no twinning. Rather often they are replaced, at least partly, by chlorite or amphibole. Frequently, they are zoned with a Mg-rich core and Fe-rich rim. With increasing size there is a tendency to radiate intergrowth with plagioclase.

Rare poikilitic pyroxene exists as phenocrysts near the border of the sill and includes pseudomorphs of chlorite after *olivine*. Other pseudomorphs of chlorite may partly have been olivine as judged by the outlines. These pseudomorphs can be found within larger batches of chlorite or enclosed by plagioclase. However, fresh olivine, mentioned by Pauly (1958) as a main constituent of the Wasenbach dolerite, does not occur.

Occasionally, pyroxene is bordered by a narrow rim of *aegirine*. More often the latter grew homoaxially into the interstitial space (Fig. 6). Individual grains of aegirine are confined to this kind of setting as well.

Strong pleochroitic *amphibole* occurs in small but rather varing amount (Fig. 4a). The unhedral grains up to 1 mm in length are situated exclusively in the interstitial space.

Fig. 5
Albitized plagioclase (chessboard albite).

Katophorite is to be found besides an amphibole of the arfvedsonite series. The latter frequently replaces the katophorite at its rim or along the cleavage (Fig. 7).

Biotite occasionally constitutes an accessory mineral. Its mostly minute crystals have xenomorphic outlines. They have grown on or in the surrounding of titanomagnetite.

Apatite is always present in an amount up to 3.8 vol-% (Fig. 4b) in typically very long euhedral needles. The maximum length measured was 1.7 mm.

From the opaque minerals *titanomagnetite* is most important constituting 5 to 13 vol-% (Fig. 4c). The euhedral crystals show mostly ilmenite exsolution. This is especially obvious where the magnetite component has been removed and just leukoxene minerals left. The opaque ilmenite exsolution lamellae stand out on (111) position of the former magnetite (Fig. 8).

Most important of the secondary minerals is *chlorite* ranging between 13 and 32 vol-% in the rock. It dominantly occurs in the interstitial space and replaces pyroxene and plagioclase. Three different chlorite types can optically be distinguished in varing amounts. The third type is mostly found as monocrystal pseudomorphs after pyroxene and perhaps olivine. All three types belong to Mg-Fe-chlorites (delessite).

Epidote is a secondary accessory mineral, its color changing due to a different Fe-content. As radially fibrous aggregates, epidote often grew into the interstitial space replacing chlorite.

Calcite and *pyrite* are secondary accessories as well, the latter constituting big, mainly idiomorphic crystals.

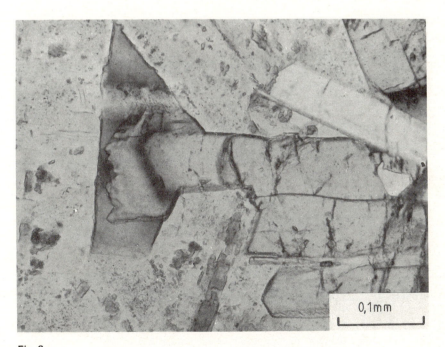

0,1mm

Fig. 6
Homoaxial growth of aegirine (zoned) on an augite into the interstitial space between plagioclase.

Fig. 7
Katophorite (grey) being replaced along its rim and its cleavage by an amphibole of
the arfvedsonitic series (black).

Fig. 8
Ilmenite exsolution lamellae on (111) of a former titanomagnetite.

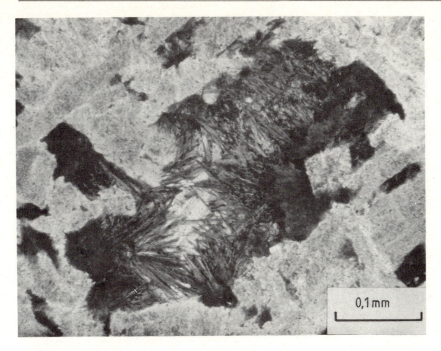

Fig. 9
Stilpnomelane needles replace calcite of the interstitial space.

Stilpnomelane occurs in rather varing amounts. Its fibrous aggregates replace chlorite, calcite (Fig. 9), pyroxene, and plagioclase.
Cross-fibrous *hornblende-asbestos* (confirmed by x-rays) occasionally fills joints within the sill and in the country rock near the contact. Frequently it is intergrown with albite.

Genesis of the Dolerite Sill

The dolerite sill of Wasenbach is part of the widespread mafic volcanism which accompanied the subsidence in the Lahn and Dill syncline area. The intrusive nature does not allow to reveal its age directly. The smaller sills in the country rock are petrographically similar to the effusive spilites of that area. These show, however, a different petrographic character from Wasenbach sill and surely belong to the meta-tholeiites described by Wedepohl et al. (1983), whereas Wasenbach sill is a meta alkali basalt. A comparison with the other mafic meta-volcanics of the Lahn and Dill synclines makes it probable that Wasenbach sill belongs to the Givetian/Adorfian phase and not to the Early Carboniferous phase. This is supported by the statement of Meyer (1981) that the Givetian/Adorfian "Schalstein" (pyroclastics) shows an element configuration of a certain alkali basaltic character.
The intrusive position of Wasenbach dolerite was controlled by one of the small sills mentioned above. This one can be followed just 1m above the upper contact of Wasenbach sill along its entire outcrop. It appears that this small sill together with its underlying

metamorphosed sediments provided the top boundary for the intrusion of the dolerite. The latter (meta alkali basalt) belongs thus to a later intrusive phase.

The melt of Wasenbach dolerite contained rather few intratelluric phenocrysts which can be found near the contact. These phenocrysts were resorbed rather soon in the interior parts of the sill as shown by the texture (Fig. 3a). Changes of the conditions of crystallisation as recognized from modal content, from grain size, and texture (Fig. 3 and 4) can be attributed to an increase of the fluid phase from the contact inwards. This demonstrates that solidification advanced about three times faster from the bottom than from the top. Fractionation processes were rather slight during crystallization, as indicated by a very small fraction of late latitic differentiate filling contraction joints. Its fine grained texture reveals that the fluid pressure dropped during the late stage contraction of the sill due to cooling.

The interrelation of the secondary minerals reveals that there were at least two separate postmagmatic alterations. Spilitization of the rock took place probably during diagenesis. The hydrothermal filling of joints and fractures might have occurred during this process or might have been a separate phase.

The rather weak metamorphism during the Hercynian orogeny can be recognized by the late formation of stilpnomelane. This mineral is otherwise very rare in the Lahn and Dill synclines but has been described by Flick (1978) from the keratophyric sill of Rupbach valley not far from Wasenbach in an even greater amount. Thus it appears that metamorphic conditions were slightly higher at the southwestern end of the Lahn syncline than elsewhere.

Discussion

Mafic volcanism in the Rhenish Mountains has been attributed to tholeiites due to geochemical investigations (Wedepohl et al. 1983). Wasenbach dolerite as described here shows significant mineralogical criteria to be associated with alkali basalts which can be recognized despite secondary spilitization. These criteria are summarized below:

Alkali pyroxene and alkali amphiboles are not just a late differentiate. They are present throughout the sill and reveal a primary surplus of Na over Al. This cannot be detected by chemical analysis because of the spilitic Na-metasomatism.

In Wasenbach dolerite there is no indication for two different pyroxenes, a Ca-rich and a Ca-poor one, being typical in tholeiites.

From its color the biotite indicates a raised Ti-content being typical for alkali rocks as well as being rather rich in apatite.

Small lumps of country rock up to 20 cm in diameter can be found occasionally being recrystallized. The argillaceous parts of the originally calcareous shales have been altered to alkali feldspar, alkali amphibole and alkali pyroxene. This alteration is identical with fenitization processes around alkali intrusions.

It can be expected that more meta alkali basaltic volcanics will be found in the Lahn-Dill syncline area. They are not common amounting to a few among a couple of hundred occurrences of mafic volcanics. So they might have been missed during sampling for the geochemical investigations. In the Frankenwald area (belonging to the Saxothuringian zone of the Hercynian mountain chain), Wirth (1984) described a great amount of meta alkali basalts besides intraplate tholeiites.

The finding of a meta alkali basalt in the southwestern Lahn syncline fits well to the record of the bulk of intraplate tholeiites in the Lahn and Dill synclines by Meyer (1981) except for three samples. These three show certain affinities to ocean ridge tholeiites.

They have been looked upon as representative for the whole Lahn-Dill syncline area by Wedepohl et al. (1983). According to the last account, there is a change from intraplate tholeiites to ocean ridge tholeiites in the Rhenish Mountains from north to south. This should indicate an increase of mantle heat flow approaching the Mid-German Crystalline Rise ("Mitteldeutsche Schwelle"). However, these three samples come from a series considered by Ahrendt et al. (1978, 1983) and Engel et al. (1983) to be part of the Gießen greywacke unit which they interpret as a nappe coming from the south of the Rhenohercynian zone. Instead of being representative of the Lahn-Dill syncline area, these samples with ocean ridge tholeiitic tendencies support the hypothesis of nappe tectonism. As a consequence, the geotectonic position of the whole mafic volcanism in the Rhenohercynian zone appears to be rather uniform, situated on continental crust without a sign of approaching oceanic environment.

Summary

Wasenbach dolerite is a sill intruded into sediments at or near the Early/Middle Devonian boundary probably during the Givetian/Adorfian phase of volcanism in the southern Lahn syncline.

Differentiation processes were rather slight during solidification which was three times faster from below than from the top. Differentiation was pronounced at a late stage leading to a very small fraction of latitic composition.

Spilitization of the rock occurred during postmagmatic alterations.

The frequent formation of late stilpnomelane in Wasenbach sill as well as in a keratophyric sill in the neighborhood can be correlated with the Hercynian metamorphism during orogenesis. This leads to the assumption of a slightly higher metamorphism in the south-western Lahn syncline than in the rest of the region.

Alkali pyroxenes and alkali amphiboles throughout the sill show a surplus of Na over Al. These observations together with other mineralogical criteria, especially fenitization of enclosed countryrock, demonstrate an alkali basaltic origin for this sill whereas solely on geochemical data all mafic volcanics of the Lahn-Dill syncline area have been assigned to tholeiitic origin.

Wasenbach sill as a meta alkali basalt fits in with meta intraplate tholeiites that are widespread in the Lahn-Dill syncline area. Together they are representative for a continental geotectonic environment.

Samples of ocean ridge tholeiitic affinities are not representative for the Lahn-Dill syncline aerea. They rather support the hypothesis of nappe tectonism in the southern Rhenish Mountains.

Acknowledgement

Microprobe analyses were gratefully provided by G. C. Amstutz and V. Stähle (Heidelberg). H. Requadt (Mainz), M. Hardi and J. Schmidt (both Heidelberg) let us use their manuscript maps of adjoining areas. The microscopic photos were made by K. Schacherl (Heidelberg). We thank H. W. Pfefferkorn (Heidelberg) for helpful criticism.

Refernces

Ahrendt, H., J. C. Hunziker and K. Weber (1978): K/Ar-Altersbestimmungen an schwachmetamorphen Gesteinen des Rheinischen Schiefergebirges. Z. dt. geol. Ges. 129, 229–247.

Ahrendt, H., N. Clauer, J. C. Hunziker and K. Weber (1983): Migration of Folding and Metamorphism in the Rheinische Schiefergebirge Deduced from K-Ar and Rb-Sr Age Determinations. In: Intracontinental Fold Belts, Martin, H. and F. W. Eder (Ed.), 323–338, Springer, Berlin–Heidelberg–New York–Tokyo.

Engel, W., W. Franke, C. Grote, K. Weber, H. Ahrendt and F. W. Eder (1983): Nappe Tectonics in the Southwestern Part of the Rheinisches Schiefergebirge. In: Intracontinental Fold Belts, Martin, H. and F. W. Eder (Ed.), 267–287, Springer, Berlin–Heidelberg–New York–Tokyo.

Flick, H. (1978): Der Keratophyr vom Rupbachtal (südliches Rheinisches Schiefergebirge). Mainzer geowiss. Mitt. 7, 77–94.

Hentschel, H. (1970): Vulkanische Gesteine. Erl. geol. Kt. Hessen 1:25 000, Bl. 5215 Dillenburg, 2. Aufl., 314–374.

Kegel, W. (1922): Abriß der Geologie der Lahnmulde. Abh. Preuß. Geol. L.-Anst., N. F. 86, 1–91.

Meyer, K. (1981): Geochemische Untersuchungen an Spiliten, Pikriten, Quarzkeratophyren und Keratophyren des Rhenoherzynikums. Diss. Göttingen, 1–121.

Pauly, E. (1958): Das Devon der südwestlichen Lahnmulde und ihrer Randgebiete. Abh. hess. L.-Amt Bodenforsch. 25, 1–138.

Requadt, H. and K. Weddige (1978): Lithostratigraphie und Conodontenfaunen der Wissenbacher Fazies und ihrer Äquivalente in der südwestlichen Lahnmulde (Rheinisches Schiefergebirge). Mainzer geowiss. Mitt. 7, 183–237.

Wedepohl, K. H., K. Meyer and G. K. Muecke (1983): Chemical Composition and Genetic Relations of Meta-Volcanic Rocks from the Rhenohercynian Belt of Northwest Germany. In: Intracontinental Fold Belts, Martin, H. and F. W. Eder (Ed.), 231–256, Springer, Berlin–Heidelberg–New York–Tokyo.

Wirth, R. (1984): Geochemie der paläozoischen Magmatite des Frankenwaldes. Pikrite-Diabase-Keratophyre. Geol. Jb. D63, 23–57.

Heat Flow and Kinematics of the Rhenish Basin

O. Oncken

Institute for Geology and Paleontology, J. W. Goethe-University, Frankfurt, Federal Republic of Germany

Key Words

Rank analysis
Paleogeothermics
Tectonic subsidence
Basin kinematics
Paleogeotectonic setting

Abstract

The rank of organic matter dispersed in sediments of the Rhenish fold belt permits the evaluation of some structural and paleogeothermal aspects of the preceeding basin. Present anticlines apparently have developed from zones of relatively lesser basin subsidence with elevated heat flow. The internal structure of the basin filling produced by this pattern of subsidence supports the view of a former basin site on continental crust. The somewhat differing picture offered by rank data from the southern Massif, however, probably is influenced by synkinematic coalification due to tectonic superposition.
The kinematics of the basin reflected in the mode of total subsidence and differential subsidence displays pronounced phases and rough synchronization throughout the basin. 'Backstripping' the sedimentary pile in order to neutralize its effect on subsidence reveals a strongly segmented mode of the purely tectonic subsidence with a rapid phase in Siegenian to Givetian time followed by a complete termination during the Upper Devonian. This, in combination with an unusually high and variable heat flow, the given pattern of subsidence, and the repeated volcanic activity suggests a past situation most closely resembling a continental back-arc region whose evolution is essentially controlled by crustal stretching and thermal processes.

Introduction

Recent discussion on the paleotectonic setting of the Rhenish Massif (Fig. 1) has essentially centered on aspects of the sedimentary and volcanic evolution as well as on style and process of orogenic deformation (Franke et al. 1978, Walliser 1981, Ziegler 1982, Weber and Behr 1983, Wedepohl et al. 1983, Giese et al. 1983). The present analysis focuses on questions related to the geosynclinal evolution of the Rhenish basin. The

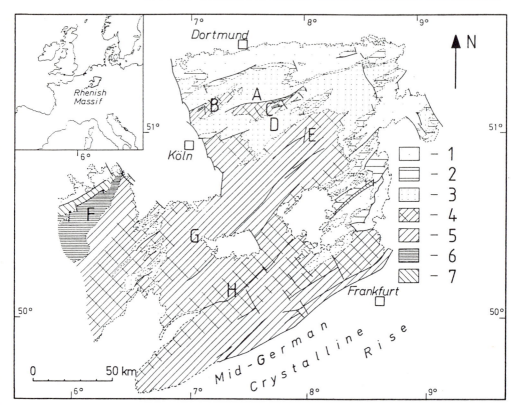

Fig. 1
Geological map of the Rhenish Massif. Numbers and capital letters refer to the following time-stratigraphic and tectonic units: **1** Carboniferous, **2** Upper Devonian, **3** Middle Devonian, **4** Emsian, **5** Gedinnian and Siegenian, **6** Cambrian to Silurian (Caledonian basement), **7** undifferentiated Devonian.
A = Lüdenscheid syncline, **B** = Remscheid anticline, **C** = Ebbe anticline, **D** = Gummersbach syncline, **E** = Müsen anticline, **F** = Venn anticline, **G** = Eifel anticline, **H** = Mosel syncline.

approach adopted proceeds on the fact that data on coal rank contain information on the former thermal state of the basin as well as on the history and the amount of basin subsidence. As is suggested by recent models of basin formation, this information would seem to be of considerable value for a paleogeotectonic interpretation.

Analysis of Coalification

The rank of organic matter dispersed in sediments is essentially dependent on the effective temperature and the time of coalification (see M. and R. Teichmüller 1954, Karweil 1955, Patteisky and Teichmüller 1960, Stach et al. 1975). The rank is thus a function of the depth of maximum burial, of the prevailing geothermal gradient, and of the burial history. The knowledge of this function (see appendix) therefore permits the evaluation of the information sought provided that coalification was stopped before folding (prekinematic coalification).

In the case of the Rhenish Massif, most authors agree on coalification being terminated before or during the beginning of folding with few exceptions in the northeastern Massif (e. g. Wolf 1972, Teichmüller et al. 1979; further exceptions in the southern Massif will be discussed). The information contained in the rank data therefore reflects a pre- to early synkinematic state of the Rhenish basin corresponding roughly to the time of maximum burial of the presently exposed strata. In contrast to other methods proposed (e. g. Buntebarth 1979, Royden et al. 1980, Stegena et al. 1981), the objective of the present approach (see appendix) is the evaluation of the named information from randomly distributed surface data of an area strongly affected by folding and erosion. This situation necessitates the construction and analysis of rank curves and gradients.

Some problems are raised concerning the accuracy of the results when applying this procedure to such a folded area. The quantitative evaluation requires a reliable knowledge of sedimentary thicknesses and of their respective stratigraphic ages. Possible error is minimised by averaging thicknesses over a greater area and by adding the sedimentary sequences of a longer time span with the consequence, however, of reduced temporal and spatial resolution. All results presented are therefore to be seen as rough approximations rather than exact figures.

Results of Analysis of Rank Curves

The map in Fig. 2 shows the rank gradients constructed from the rank profiles of 56 sections of the Rhenish Massif. Apart from an unexpected high variability, there is a distinct relationship between the rank gradients and large scale tectonic structures. Anticlines obviously exhibit higher gradients than the neighboring synclines. These fairly constant differences cannot be attributed to the stratigraphic age of the involved strata nor to their lithology, since neither significantly affects the correlation shown.

With the help of these rank gradients, the thickness of the formerly superimposed strata – corresponding to the amount of erosion – can be evaluated. As was to be expected, most erosion has occurred on top of the anticlines (3–6 km). However, considerable denudation has also affected the synclines (1–4 km) although they still contain sediments as young as Upper Devonian to Lower Carboniferous. As is suggested by the comparative study of the shape of the rank curves and of the course of rank isolines in relation to stratigraphic age (cf. also Paproth and Wolf 1973), the filling of the evolving synclines probably had continued while sedimentation on the neighboring anticlines – with exception of the northermost anticlines – had already stopped. North-south striking zones (the Eifel depression in particular) show a slightly different picture with small rank gradients and only minor uplift and erosion ($\leqslant 1$ km).

The interpretation of rank data in the southern Massif meets with other problems. All strata exhibit roughly the same rank (see Wolf 1978) in spite of their different stratigraphic ages from Siegenian to Upper Devonian, therefore inhibiting the construction of a reliable rank curve with the exception of two smaller areas. This feature suggests continuing coalification during folding (synkinematic coalification) which caused a leveling-out of earlier rank differences. Another effect of this process is the conversion of the original – prekinematic – rank curves to curves with a flatter gradient which, if they can be constructed at all, give values for the superimposition of the youngest exposed strata which are too high. Furthermore, the geothermal gradients calculated are lower than the original ones.

Although this apparent superimposition with a fairly constant 3–4 km for sediments of a wide stratigraphic range definitely overrates the true maximum burial of the youngest presently exposed sediments, the total superimposition cannot have been much lower

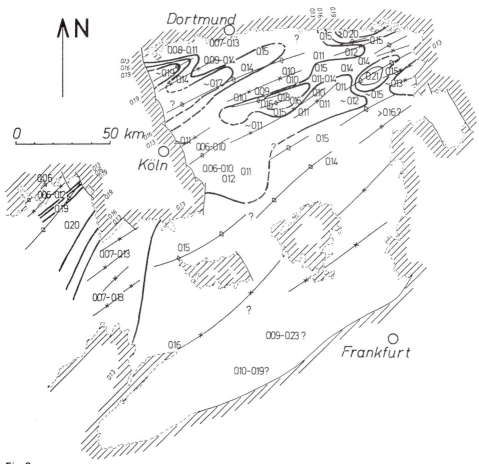

Fig. 2

Gradients of maximum vitrinite reflectance (in % Rmax % Rmax/100 m between 2.5 and 5.5 % Rmax)
in the Rhenish Massif; heavy solid lines: isogradients verified by data; broken lines: assumed course of
isogradients (source of rank and thickness data is listed in Oncken 1984).

as may be inferred from the rather high rank of 5–7 % Rmax (Wolf 1978). On the whole,
these observations could easily be explained if part of the superimposition had been tectonic in nature. This would imply that thin-skinned nappe tectonics ($\leqslant 3$ km) during
folding – with the involved sediments now eroded – had supported continuing, synkinematic coalification and the leveling-out of rank differences in the underlying sediments in parts of the southern Rhenish Massif (cf. Weber and Behr 1983).
The thickness of the underlying sediments can be added to the amount of erosion reconstructed to give the total thickness of the geosynclinal pile. This is easily done in
some of the anticlines where the pre-Devonian basement is exposed. For the synclines,
only values for part of the total thickness are ascertainable. However, the study of the
evolution of rates of sedimentation (cf. van Hinte 1978) gives an idea of the probable
order of magnitude of total thickness reached in the synclines (see Fig. 3). Since the

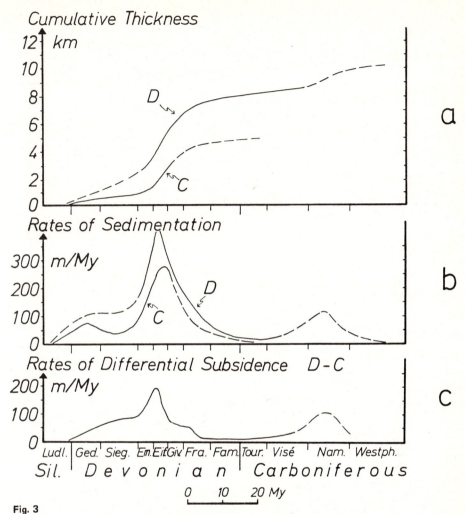

Fig. 3

Subsidence history of the Ebbe anticline (C) and the Gummersbach syncline (D). Solid lines depict observed thickness data; broken lines represent computed thickness and subsidence data as inferred from rank analysis and from extrapolating curves for rates of sedimentation backwards. The lower broken part of curve (D) in (a) is thus calculated from extrapolation of curve (D) in (b). Fig. 2c traces the history of differential subsidence between these two neighboring structures. Solid-lined curves in (b) and (c) are constructed from data points in (a) in intervals of 5 mys. Time scale is chosen according to van Eysinga (1975).

rates seem to follow an identical trend wherever they are documented, the rates of sedimentation of the synclines can be extrapolated backwards by analogy to their known equivalents in the neighboring anticlines.

The isopach map constructed on this basis (Fig. 4) exhibits minimum subsidence to be related to the present anticlines (cf. Meyer and Stets 1980). Part (i. e. about 1–3 km) of the filling of the synclines, however, is due to the Carboniferous redistribution of sediments during folding and the synkinematic flysch phase when sedimentation on most anticlines had already stopped (see above). In spite of the orogenic deformation, the present tectonic structure thus clearly traces the pattern of basin subsidence with the possible exception, however, of the southern Massif which is affected by a more complex orogenic overprinting. The influence of north-south striking zones with their average subsidence on the present tectonic picture is caused, on the other hand, by their relatively limited uplift indicating their reduced mobility.

It appears that the geosynclinal evolution of these structures does not significantly affect facies and sedimentation (Franke et al. 1978). They are therefore to be seen as structures within the basin filling, evolving by differing rates of subsidence without strong paleo-geographic influence because of a positive balance between the rate of clastic supply and the rates of subsidence and of differential subsidence.

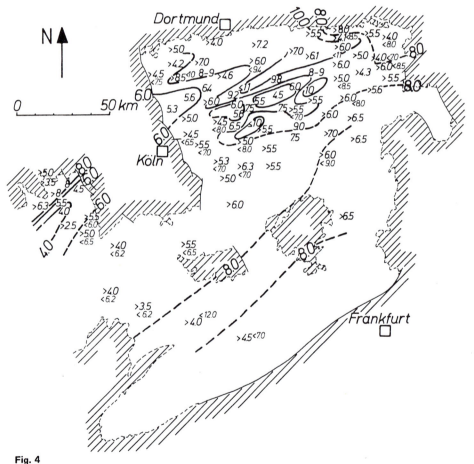

Fig. 4

Total burial of pre-Devonian basement in km. Data represent approximate values. Where data are not obtainable, minimum observed thickness is given; smaller-sized accompanying number indicates the probable upper limit for the estimated total thickness as inferred from extrapolation of known rates of sedimentation. Heavy solid lines delineate isopachs; when broken they show their assumed course.

The palinspastically restored subsidence pattern, it may be noted, shows the same ampli-
tudes (1–4 km, excluding the flysch phase) and half-wavelengths (10–60 km) of the
basement topography which has been found to be characteristic for the buried basement
structure of rifted continental crust. This observation thus strongly supports the interpretation of the crustal position of the Rhenish basin derived from analysis of the basin
volcanics (Wedepohl et al. 1983) and from seismic analysis (Giese et al. 1983).

The conversion of rank gradients to paleogeothermal gradients (with the help of equation (5), see appendix) reveals a high geothermal variability similar to the pattern of
rank gradients (cf. Figs. 2 and 5). Moreover, the geothermal gradients reach fairly high
values ranging between 30 and more than 80 °C/km with an average of 50 °C/km. Assuming a heat conductivity of 2–2.5 W/m °K this corresponds to a rather high mean heat
flow of 115 mW/m². Although the differing burial histories of the anticlines and synclines are incorporated in the calculation, anticlines still show considerably higher thermal
gradients than do the synclines.

North-south striking zones again somewhat depart from the general picture with their
rather low gradients. This is particularly the case in the Eifel depression and the southern

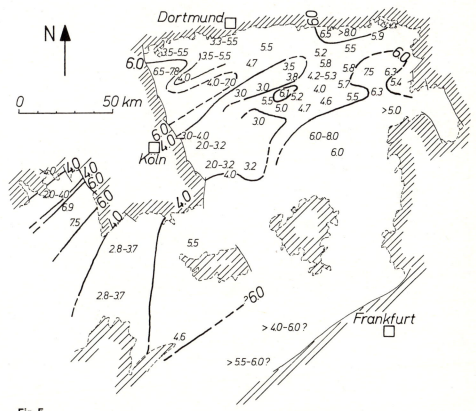

Fig. 5
Paleogeothermal gradients in the Rhenish basin in °C/100 m. Heavy solid lines depict lines of equal
paleogeothermal gradients; broken lines represent their probable course. Gradients in the southern
Massif are minimum values (see text). The gradients are computed from gradients of coalification;
for the method, see Oncken (1984).

Bergisches Land. Exceptionally high gradients are met north of the Warstein anticline; they probably are related to the late-Variscan pluton of the 'Lippstädter Gewölbe' (Wolf 1972, Paproth and Wolf 1973, Teichmüller et al. 1979) thus reflecting a predominantly syn- to postkinematic heat flow regime in this area.

The interpretation of the heat flow picture in other parts of the Rhenish Massif is impeded by a possibly complex interaction of several causes. As mentioned above, in most cases the present picture of coalification can be seen to reflect a predominantly pre-kinematic thermal state of the basin. The equations describing the relationship moreover stress the role of the progress of subsidence expressed as the integral of the burial history, thereby including the integral of the temperature history of the sediment. The gradients computed therefore mainly represent the mean effective thermal state of the late geosynclinal stage rather than the peak gradients reached, especially if these are short-lived.

Temporal changes of heat flow are certainly contained in the paleogeothermal picture. In order to contribute to this picture, the uplift of the anticlines should coincide with the thermal climax while the still-subsiding synclines should retain a later cooler stage. This issue, being strongly dependent on the heat flow properties of the basin filling and on the rate of subcrustal changes in thermal state, is however thought to be of minor importance.

The ultimate reason for the differentiated heat flow field has to be sought in the answer to the question of whether the differential subsidence of the basin (Fig. 4) is caused by a differential subcrustal temperature field or whether the differentially subsiding basement with its presumably higher conductivity is disturbing a primarily homogeneous temperature field. Both effects might create a similar picture, and both would be able to explain the relation between paleogeothermal gradients and total subsidence shown in Fig. 6.

The latter assumption, however, would not be able to explain what was the principle cause of the differential subsidence, if isostatic balance was to be maintained on a million-year (my) scale (although the geometry of the fault-controlled blocks might contribute to isostatic differential subsidence). Investigations of the ocean floor have shown that its subsidence is strongly controlled by heat loss during spreading and ageing (e. g. Sleep

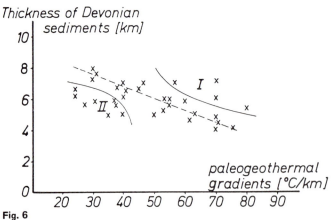

Fig. 6

Relation of total thickness of Devonian sediments to paleogeothermal gradients in the Rhenish Massif. Roman numbers refer to the following areas: I. northeastern Massif (Warstein anticline, Nuttlar syncline, East Sauerland anticline); II. Eifel depression, southern Bergisches Land and Inde syncline.

1971), thus suggesting that the smaller subsidence of the anticlines might have been effected by their higher temperatures. Finally, convective heat transfer by pore fluids probably has affected the temperature field, especially if concentrated on the slower subsiding structures being supported by growth faults and/or lithology.

The figures from the southern Massif only represent apparent paleogeothermal gradients based on rank data obviously altered by synkinematic coalification (see above), thus giving the lower limit of the original gradients which presumably were somewhat higher (cf. estimates of Weber and Behr 1983).

The relationship between paleogeothermal gradients and basin thicknesses (the thicknesses of the Carboniferous flysch is not considered because it is mainly related to orogenic processes) is reemphasized by the diagram in Fig. 6, notwithstanding the ultimate causal connection. In spite of some scatter which probably is due to error as well as further controlling factors such as intralithospheric inhomogeneities, the diagram shows the linear relationship as well as the above-mentioned exceptions: the 'cool' north-south striking zones and the 'hot' northeastern Massif.

Basin Kinematics and Tectonic Subsidence

The quantitative evaluation of the progress of subsidence gives further information on the kinematics of the Rhenish basin with its characteristic aspects being illustrated by the example shown in Fig. 3. Since sedimentation is assumed to have proceeded under predominantly shallow marine conditions (e.g. Franke et al. 1978, Walliser 1981), rates of sedimentation may be viewed as rates of subsidence to a first approximation. These rates, compiled from different areas of the Rhenish Massif (see Fig. 7), display the same well defined phases of subsidence which have already been interpreted by Franke et al. (1978) as the succession of the 'Caledonian Molasse', a phase of stagnation, and a final flysch phase. Obviously, the evolution of the rates of sedimentation is roughly synchronized throughout the former basin. This is equally valid for the course of the differential subsidence between neighboring areas. Being concentrated in the Emsian and Middle Devonian and declining almost totally in Late Devonian, the effects of differential subsidence are best documented in the oldest part of the basin filling.

The second peak of the rates in the Carboniferous is related to the climax of flysch sedimentation during the orogenic stage which apparently reactivated the structural pattern of the basin filling.

However, the mobility of the basin floor expressed in these curves is not entirely synchronized throughout the basin. There is a slight delay from south to north similar to the shift of successive depo-centers established by Kegel (1950). The time of the maximum mobility of the respective areas shifts northwards — not perpendicular to the present strike of the fold belt — from Siegenian to Mid-Devonian time with the exception of the southern Massif (see Fig. 8). Both the amount of subsidence and the time of maximum mobility of north-south trending zones appear to lag somewhat behind the adjacent regions and to disturb the northward migration of basin mobility.

It has been suggested more than once that the specific mode of subsidence in the Rhenish basin with its 3 to 4 subphases is essentially controlled by the amount of sedimentary supply and loading. The method of 'backstripping' the sedimentary pile (Sleep 1971, Watts and Ryan 1976, Bond and Kominz 1984), in order to reconstruct the purely endogenetically driven or tectonic contribution to total subsidence, yields other results.

The method proceeds by removing the total sedimentary pile accumulated over a given period since the beginning of sedimentation and by replacing it with water. Initially, this pile has to be decompacted to give the original thickness. The latter is restored

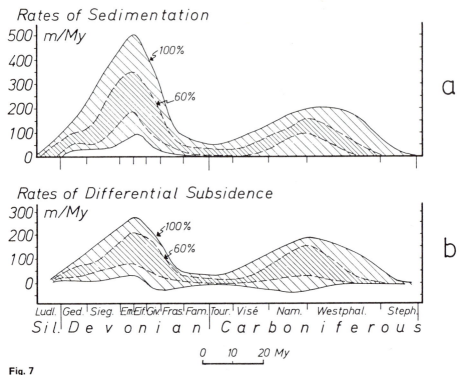

Fig. 7

Subsidence in the Rhenish basin. The entire shaded area comprises all of 39 curves of rates of sedimentation (in a) and 23 curves of rates of differential subsidence between neighboring areas (in b). The darker shaded area includes 60 % of the curves in each case. Time scale according to van Eysinga (1975).

through the ratio of the present mean density to the approximate mean density immediately after deposition — both being a function of the thickness — with the help of the density and compaction curves published by Rieke and Chilingarian (1974; see also mathematical modeling of Bond and Kominz 1984). Having performed this procedure for several time intervals (steps of 5 mys from beginning of sedimentation in this case), one obtains a corrected curve of sedimentary accumulation.

After removing the sedimentary accumulation as calculated step-by-step including the assumed depth of water, the isostatic rebound of the basement top can be computed (equations are given by Watts and Ryan 1976, and Bond and Kominz 1984). The method of calculation is based on the simple model of Airy-isostasy which treats the basement as dissected into separately moving rock prisms. This view, as against the model of flexural loading of a rigid crust, is supported by the established fault-controlled subsidence in the Rhenish basin (cf. Ziegler 1982, and Engel et al. 1983). The final result is a curve tracing the purely tectonic subsidence of the basement under isostatic conditions; the effect of crustal loading, whether sedimentary or orogenic in nature, is neutralized.

According to Sleep (1971), rifted continental crust will exhibit exponentially decaying subsidence with a time constant of ~ 50 mys. When plotted against the square root of time, an almost linear curve should result whose inclination is dependent on the amount

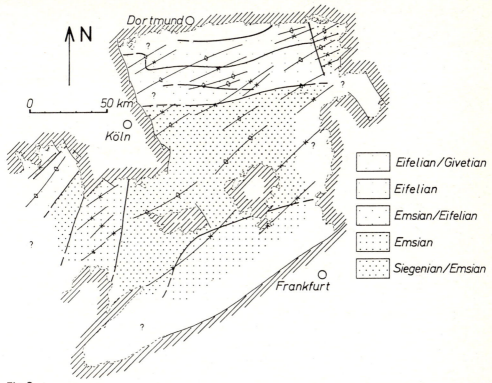

Fig. 8

Shifting of the time span of maximum rates of subsidence in the Rhenish basin during the Devonian (flysch stage excluded).

of crustal stretching (McKenzie 1978, see Fig. 9). However, a comparison of the tectonic subsidence curves from several areas of the Rhenish basin to the subsidence of stretched passive margins does not reveal a close resemblance.

Taking into account the error within reasonable limits in time scale and in stratigraphic data does not significantly affect the illustrated trends which practically remain the same. This applies as well to an error in the estimated depth of water as long as this is misjudged by less than a few hundred meters (assumed water depths are taken to remain below 200 meters until the end of the Middle Devonian and to rise to 500 m at most at the beginning of the Lower Carboniferous, declining subsequently; possible eustatic sea-level changes are not considered).

Both Devonian changes in the mode of subsidence are therefore to be viewed as true changes of lithospheric behavior caused by infra- or subcrustal processes. The rather variable basement subsidence established has thus not been controlled by a varying amount of clastic influx, the influence of which is restricted to a multiplying factor. Instead, it affected a larger area with a consequent decrease of the gradient. Furthermore, differential subsidence is also controlled tectonically, practically being confined to the time span from Siegenian to Givetian as may be seen by comparing the tectonic subsidence curves of present anticlines and synclines.

Tectonic subsidence, moreover, is virtually stopped during the Late Devonian after a phase of very rapid subsidence in the Emsian to Givetian. Further subsidence, as reflected in particular in the thick Carboniferous flysch series, must thereafter have been controlled by other factors. These include a rise in the eustatic sea level and/or raising of the base-level as well as loading of the crust by an evolving mountain belt during the Carboniferous.

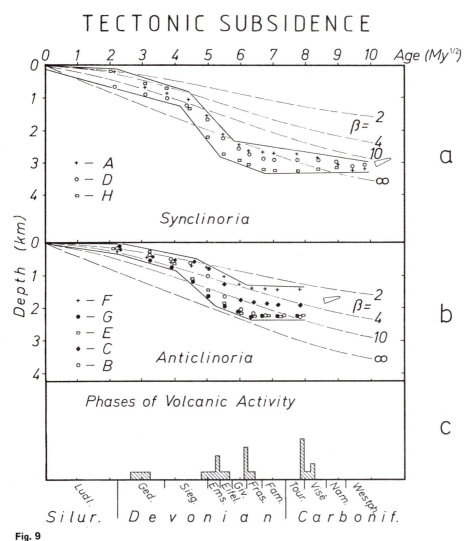

Fig. 9
Tectonic subsidence and volcanic phases in the Rhenish basin. Capital letters indicating the regions involved refer to the same areas recorded in Fig. 1. The thin broken lines in (a) and (b) trace the typical subsidence curve of crust depending on the stretching factor β (from McKenzie, 1978). Volcanic phases in (c) are displayed in a highly schematic manner not true to scale. Time scale from van Eysinga (1975) is converted to exponential form starting with the uppermost Ludlow.

Another noticeable fact highlighted by Fig. 9 is the apparently complex relationship between volcanic phases and changes in the mode of subsidence. The Lower Devonian keratophyric volcanism seems to have occurred after an acceleration of tectonic subsidence, whereas the predominantly basaltic-tholeiitic volcanism of the later two volcanic phases has been associated with a marked decrease or even with partial reversal of tectonic movement.

Discussion and Conclusions

This last observation is in good agreement with the results by Wedepohl et al. (1983) on geochemical features of the basin volcanics. These indicate a growing amount of upper mantle melting which in turn implies rising temperatures throughout basin evolution (cf. similar view of Weber and Behr 1983). According to the model of thermal contraction of the crust as an essential driving force for basin formation (Sleep 1971), this would explain the rapid decline in tectonic subsidence during the Late Devonian. As suggested by McKenzie (1978), crustal necking as a primary cause for tectonic subsidence might point to stretching and rifting associated with the start and the acceleration of tectonic subsidence and differential subsidence. However, the theoretical extent of crustal stretching needed to account for the total amount of subsidence (greater than 4!) has not been established in the Rhenish basin. Additional mechanisms might also include the hypothesis of migrating phase boundaries (e. g. Artyushkov and Sobolev 1982) which, if significant, might throw some light on the relation between the patterns of subsidence and heat flow.

Although the possible initiating and driving factors for the formation of the Rhenish basin remain somewhat uncertain and hypothetical, the results submitted put constraints on the interpretation of the paleogeotectonic setting of the basin. The pattern of differential subsidence exhibits the same basement topography found in rifted continental crust of passive continental margins and of subduction-related continental back-arc basins (e. g. Hinz et al. 1979, Bott 1979, Anderson et al. 1983). There is, moreover, the complex progress of tectonic subsidence which is rather uncommon for a passive atlantic-type margin (cf. Watts and Ryan 1976) but comparable to that of intracontinental basins (see Sleep 1971). Finally, the unusually high and variable basin heat flow and the repeated volcanic activity — equally uncommon in a passive or active continental margin — may both be reconciled with the interpretation of a back-arc region (see Brooks et al. 1984).

In agreement with the similar view of Ziegler (1982), a uniformitarian comparison of the stated observations on the Rhenish basin with present geotectonic settings thus points to a situation most closely resembling a back-arc region which, however, has not reached the stage of producing oceanic crust.

Appendix

The procedure (see Oncken, 1984, for details) requires the construction of a coalification profile by plotting the rank (given by the maximum reflectance of vitrinite) against a section of the thicknesses of the exposed strata. The trend of the resulting curve then yields the approximate depth of burial of the involved strata with the gradient of the curve being related to the former geothermal gradient. The rank gradient needed for computation is constructed from the range between 2.5 and 5.5 % maximum reflectance (Rmax). Most of the roughly 800 rank data available from the Rhenish Massif group are in this range. The lower boundary value is also defined by the behavior of organic matter

during further coalification (growing 'pleochroism' of vitrinite reflectance due to processes induced by rising superimposed load, Teichmüller et al. 1979). Moreover, the influence of burial time and of different geothermal gradients makes itself felt most strongly in just this range with differences thus being resolved more easily.
Karweil (1955) and Bostick (1971) have determined empirically the relationship between temperature, time, and vitrinite rank. Their results have been converted for the purpose of the analysis of rank curves. This is illustrated in Fig. 10 which, moreover, supports the classical view that the rank is a simple function of time and temperature which — in the case of rank gradients — can be expressed as follows:

$$\text{grad}\,T = k \cdot \text{grad}\,R\text{max} \tag{1}$$

The factor k being a function of the time of coalification can be determined from Fig. 10:

$$k = \frac{380}{t_0 - 8} + 15.5 \tag{2}$$

Equation (2) may also be calibrated to any other rank section than that chosen or to mean or minimum vitrinite reflectance curves respectively, if the relation to Rmax is known (see Teichmüller et al. 1979).
Both equations add up to:

$$\text{grad}\,T = \text{grad}\,R\text{max} \cdot \left(\frac{380}{t_0 - 8} + 15.5 \right) \tag{3}$$

where t_0 is given in mys, grad T in °C/100 m, and grad Rmax in % Rmax/100 m between 2.5 and 5.5 % Rmax.
However, this procedure does not consider a complicated burial and thus thermal history, since it assumes constant temperatures from the beginning of coalification. The cor-

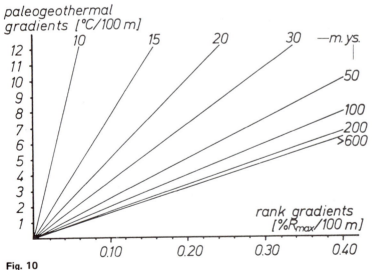

Fig. 10
Relation between paleogeothermal gradients and gradients of maximum vitrinite reflectance between 2.5 and 5.5 % Rmax considering different burial times (converted from data of Karweil, 1955, and Bostick, 1971).

rection of this time in order to obtain the effective time of coalification follows a path similar to that suggested by Bostick (1971). The burial history, inferred from sedimentary information or other sources, is determined as shown in Fig. 3a. Assuming approximately constant thermal conditions, the time t_0 has to be multiplied by the integral of the burial history x_i:

$$x_i = \frac{\int_0^{t_i} z(t)\, dt}{z_i \cdot t_i} \tag{4}$$

where the maximum depth z_i of a layer is given in km, and the time t_i from the beginning until the end of coalification of this layer is given in mys. Strictly speaking, however, this correcting factor can only be applied to a point with well defined subsidence data opposite to the intention of considering an interval of burial and coalification time as reflected in a given rank section.

The procedure is therefore simplified by referring to the mean depth z_m and time of coalification t_m of the rank section under consideration, equivalent to the stratigraphic horizon at $Rmax = 4\% (= (2.5 + 5.5)/2$; see Oncken, 1984, for simple determination of the integral of burial history).

The paleogeothermal gradients are thus computed from:

$$\text{grad}\,T = \text{grad}\,Rmax \cdot \left(\frac{380}{\left(\dfrac{\int_0^{t_m} z(t)\, dt}{z_m \cdot t_m} \right) \cdot t_m - 8} + 15.5 \right) \tag{5}$$

The interpretation of the results has to consider, however, that this computed paleogeothermal gradient does not necessarily reflect the maximum gradient or the gradient at the end of burial. Rather, the equation gives the mean effective thermal state with emphasis on the final geosynclinal stage, since it implicitly includes the integral of the thermal history of the sediment. Temporal variations of heat flow cannot be resolved by rank data alone.

References

Anderson, R. E., Zoback, M. L., and Thompson, G. A. (1983): Implications of selected subsurface data on the structural form and evolution of some basins in the northern Basin and Range province, Nevada and Utah. Bull. Geol. Soc. Amer. 94, 1055—1072.

Artyushkov, E. V. and Sobolev, S. V. (1982): Mechanism of Passive Margin and Island Seas Formation. In: Watkins, J. S. and Drake, C. L. (eds.), Studies in Continental Margin Geology. Am. Ass. Petr. Geol. Mem. 34, 689—701.

Bond, G. C. and Kominz, M. A. (1984): Construction of Tectonic Subsidence Curves for the Early Paleozoic Miogeocline, southern Canadian Rocky Mountains: Implications for Subsidence Mechanisms, Age of Breakup and Crustal Thinning. Bull. Geol. Soc. Amer. 95, 155—173.

Bostick, N. H. (1971): Time as a factor in thermal metamorphism in phytoclats. In: 7ème Congr. Int. de Stratigr. et de Geol. du Carbonifère, Compte Rendu 2, 183—192.

Bott, M. H. P. (1979): Subsidence mechanisms at passive continental margins. Am. Ass. Petr. Geol. Mem. 29, 3—10.

Brooks, D. A., Carlson, R. L., Harry, D. L., Mella, P. J., Moore, R. P., Rayhorn, J. E., and Tubb, S. G. (1984): Characteristics of Back Arc Regions. Tectonophysics 102, 1—16.

Buntebarth, G. (1979): Eine empirische Methode zur Berechnung von palaeogeothermischen Gradienten aus dem Inkohlungsgrad organischer Einlagerungen in Sedimentgesteinen mit Anwendung auf den mittleren Oberrhein-Graben. Fortschr. Geol. Rheinld. u. Westf. 27, 97—108, Krefeld.

Engel, W., Franke, W., and Langenstrassen, F. (1983): Paleozoic Sedimentation in the Northern Branch of the Mid-European Variscides. Essay of an Interpretation. In: H. Martin and W. Eder

(Editors), Intracontinental Fold Belts. Case Studies in the Variscan Belt of Europe and the Damara Belt in Namibia. Springer, Berlin Heidelberg New York Tokyo, pp. 9—41.

Franke, W., Eder, W., Engel, W., and Langenstrassen, F. (1978): Main Aspects of Geosynclinal Sedimentation in the Rhenohercynian Zone. Zt. dt. geol. Ges. 129, 201—216.

Giese, P., Joedicke, H., Prodehl, C., and Weber, K. (1983): The Crustal Structure of the Hercynian Mountain System — A Model for Crustal Thickening by Stacking. In: H. Martin and F. W. Eder (Editors), Intracontinental Fold Belts. Case Studies in the Variscan Belt of Europe and the Damara Belt of Namibia. Springer, Berlin Heidelberg New York Tokyo, pp. 405—426.

Hinz, K., Schlueter, H. U., Grant, A. C., Srivastava, S. P., Umpleby, D., and Woodside, J. (1979): Geophysical Transsects of the Labrador Sea: Labrador to Southwest Greenland. Tectonophysics 59, 151—183.

Karweil, J. (1955): Die Metamorphose der Kohlen vom Standpunkt der physikalischen Chemie. Zt. dt. geol. Ges. 107, 132—139.

Kegel, W. (1950): Sedimentation und Tektonik in der Rheinischen Geosynklinale. Zt. dt. geol. Ges. 100, 267—289.

McKenzie, D. (1978): Some Remarks on the Development of Sedimentary Basins. Earth Planet. Sci. Lett. 40, 25—32.

Meyer, W. and Stets, J. (1980): Zur Palaeogeographie von Unter- und Mitteldevon im westlichen Schiefergebirge. Zt. dt. geol. Ges. 131, 725—751.

Oncken, O. (1984): Zusammenhänge in der Strukturgenese des Rheinischen Schiefergebirges. Geol. Rdsch. 73, 619—649.

Paproth, E. and Wolf, M. (1973): Zur palaeogeographischen Deutung der Inkohlung im Devon und Karbon des nördlichen Rheinischen Schiefergebirges. N. Jb. Geol. Paleont. Mh. 1973, 469—493.

Patteisky, K. and Teichmüller, M. (1960): Inkohlungsverlauf, Inkohlungsmaßstäbe und Klassifikation der Kohlen auf Grund von Vitrit-Analysen. Reprint from „Brennstoff-Chemie" 3 (v 41), Essen.

Rieke, H. H. and Chilingarian, G. V. (1974): Compaction of Argillaceous Sediments. Developments of Sedimentology 16, 424 pp. Elsevier, Amsterdam London New York.

Royden, L., Slater, J. G., and von Herzen, J. P. (1980): Continental Margin Subsidence and Heat Flow: Important Parameters in Formation of Petroleum Hydrocarbons. Am. Assoc. Petr. Geol. Bull. 64, 173—187.

Sleep, N. H. (1971): Thermal Effects of the Formation of Atlantic Continental Margins by Continental Break-up. Geophys. J. Roy. astr. Soc. 24, 325—350.

Stach, E., Mackowsky, M.-Th., Teichmüller, M., Taylor, G. H., Chandra, D., and Teichmüller, R. (1975): Stach's Textbook of Coal Petrology. 2nd ed., 428 pp., Berlin, Stuttgart.

Stegena, L., Horváth, F., Sclater, J. G., and Royden, L. (1981): Determination of Paleotemperature by Vitrinite Reflectance Data. Earth Evolution Sci. 3—4/1981, 292—300.

Teichmüller, M. and Teichmüller, R. (1954): Die stoffliche und strukturelle Metamorphose der Kohle. Geol. Rdsch. 42, 31 pp.

Teichmüller, M., Teichmüller, R., and Weber, K. (1979): Inkohlung und Illitkristallinität. Vergleichende Untersuchungen im Mesozoikum und Palaeozoikum von Westfalen. Fortschr. Geol. Rheinld. u. Westf. 27, 201—276, Krefeld.

van Eysinga, F. W. B. (1975): Geological Time Table. 3rd Edition, Elsevier, Amsterdam.

van Hinte, J. E. (1978): Geohistory Analysis — Application of Micropaleontology in Exploration Geology. Am. Ass. Petr. Geol. Bull. 62, 201—222.

Walliser, O. (1981): The geosynclinal development of the Rheinische Schiefergebirge (Rhenohercynian Zone of the Variscides; Germany). Geol. en Mijnbouw 60, 89—96.

Watts, A. B. and Ryan, W. B. F. (1976): Flexure of the Lithosphere and Continental Margin Basins. Tectonophysics 36, 25—44.

Weber, K. and Behr, H.-J. (1983): Geodynamic Interpretation of the Mid-european Variscides. In: H. Martin and F. W. Eder (Editors), Intracontinental Fold Belts. Case Studies in the Variscan Belt of Europe and the Damara Belt of Namibia. Springer, Berlin Heidelberg New York Tokyo, pp. 427—469.

Wedepohl, K. H., Meyer, K., and Muecke, G. K. (1983): Chemical Composition and Genetic Relations of Meta-volcanic Rocks from the Rhenohercynian Belt of Northwest Germany. In: H. Martin and F. W. Eder (Editors), Intracontinental Fold Belts. Case Studies in the Variscan Belt of Europe and the Damara Belt of Namibia. Springer, Berlin Heidelberg New York Tokyo, pp. 231—256.

Wolf, M. (1972): Beziehungen zwischen Inkohlung und Geotektonik im nördlichen Rheinischen Schiefergebirge. N. Jb. Geol. Palaeont. Abh. 141, 222—257.

Wolf, M. (1978): Inkohlungsuntersuchungen im Hunsrück (Rheinisches Schiefergebirge). Zt. dt. geol. Ges. 129, 217—227.

Ziegler, P. A. (1982): Geological Atlas of Western and Central Europe. Shell Internat. Petrol., Maatschappij B. V., 110 pp., 40 enclos.

Comparison of Magnetic, Mica and Reduction Spot Fabrics in the Rocroi Massif — Ardennes, France

J. S. Rathore*

Department of Geophysics and Planetary Physics, School of Physics, The University of Newcastle-U-Tyne, U.K.

H. Hugon**

Institut de Géologie, Université de Rennes, 35042 Rennes Cedex, France

Key Words

Tectonic fabric
Hercynian
Ardennes (France)

Abstract

The magnetic, mica and reduction spot fabrics of 33 sites, along a profile in the Vallée de la Meuse, were determined to study the regional Hercynian tectonic fabric. The fabric ellipsoids are dominantly oblate with high ellipticities indicative of strong compressional strains. The dominant compression direction from both the magnetic and mica fabric is observed to be NNW-SSE. The lack of parasitic directions indicates the complete overprinting of pre-existing fabrics and the absence of the later tectonic influences. A detailed comparison of the anisotropy ellipsoid data and the strain ellipsoid data shows that their mean axial ratios are related by an empirical power relationship of the type

$$\left(\frac{\chi_i}{\chi_j}\right) = \left(\frac{\ell_i}{\ell_j}\right)^a$$

(for i = 1, 2, 3; j = 1, 2, 3 and i ≠ j) where χ_i and χ_j are orthogonal principal susceptibility axes and ℓ_i and ℓ_j are the corresponding orthogonal principal strain axes. The exponent 'a' for the sites from this study is 0.071 ± 0.005. Some implications of the relationship are discussed.

* Now at IKU A/S (Continental Shelf and Petroleum Technology Research Institute), Petroleum Technology Department, P. O. Box 1883, Jarlesletta N 7001 Trondheim, Norway.
** Now at The Department of Geology, The University of Toronto, Toronto, Canada, M5S-1E1.

Introduction

Study Background and Aims

A close association between the magnetic fabrics and petrofabrics of rock samples has been known for a long time (Graham, 1954) and the use of magnetic susceptibility anisotropy (MSA) method in structural studies has often been proposed (see the comprehensive bibliography given by Hrouda, 1982). In this respect, magnetic fabrics of all rock types have been studied and comparison studies between the magnetic and petrofabrics have been conducted to show the general co-axiality of the two fabric ellipsoids. In studies of metamorphic rock fabrics it has been shown that the degree of anisotropy is generally higher than that of sedimentary bedding fabrics, with planar orientations of magnetic foliation planes parallel to any observed cleavages (Stacey, 1960; Stone, 1963; Janak, 1965; Hrouda and Janak, 1976; Hrouda et al., 1971, 1978; Hrouda 1976, 1982; Kligfield et al., 1977, 1981). The applicability of the MSA technique to the study of structural trends has been demonstrated in a variety of environments, large and small, uniform and complex (Fuller, 1960; Girdler, 1961; Heller, 1973; Henry, 1975; Hrouda et al., 1978; Rathore et al., 1977, 1983; Rathore and Heinz, 1979, 1980; Rathore and Becke, 1980, 1983; Rathore and Mauritsch, 1983; and many others, see bibliography of Hrouda, 1982).

To date, however, only a few correlations have been carried out between the magnitudes of the strain ellipsoid axes and the anisotropic susceptibilities. In some studies it has been found that the axial ratios of the magnetic susceptibility ellipsoid are connected via a power relationship to the axial ratios of the strain ellipsoid (Wood et al., 1976; Kneen, 1976; Rathore, 1979, 1980; Rathore and Henry, 1982; Rathore et al., 1983), indicating the possibility of obtaining complete petrofabric ellipsoid data from the magnetic susceptibility ellipsoid which can be obtained quickly and reliably (ca. 2 min per sample measuring time). This paper reports on a three-fold fabric study conducted on slate samples from the Rocroi Massif in the Ardennes (France). The fabrics studied were the MSA fabrics, the mica fabrics and, where found, the reduction spot strain fabrics. It has been possible to establish relationships between the fabric parameters for the three fabrics. In the case of the mica to reduction spot strain fabrics, a 1:1 relationship was found between the fabric parameters. In the case of the MSA comparison with the other two fabrics, a power-law relationship was found.

Geological Setting and Structural Background

The deformed Cambrian series of the Rocroi Massif (Ardennes, Fig. 1) out-crops within the north-western European portion of the Hercynian Orogenic Belt. The Rocroi Massif constitutes the core of a major E. W. anticline forming the southern part of the Ardennes Massif. The general geological characteristics of the Rocroi Massif presented below are taken from the most recent comprehensive study of this massif (Beugnies, 1963). The flysh-type series of the Cambrian of the Rocroi massif are composed of alternating pelitic schists and sandstones, and are divided into Lower Cambrian, i.e. the Devillian, and Upper Cambrian, i.e. the Revinian. The Devonian series, lying unconformably on the Cambrian series, begins with a Gedinian conglomerate that grades upward into pelitic schists and thick massive sandstones. Siegenian sediments composed of thick pelitic sandstones and limestones follow in conformity on Gedinian sediments. Both Cambrian and Devonian series show northward overturned folds with E.W. subhorizontal axes. A penetrative fabric expressed as a slaty cleavage affects both Cambrian and Devonian series. Folds in the Cambrian series are of tight to isoclinal type and of open to tight

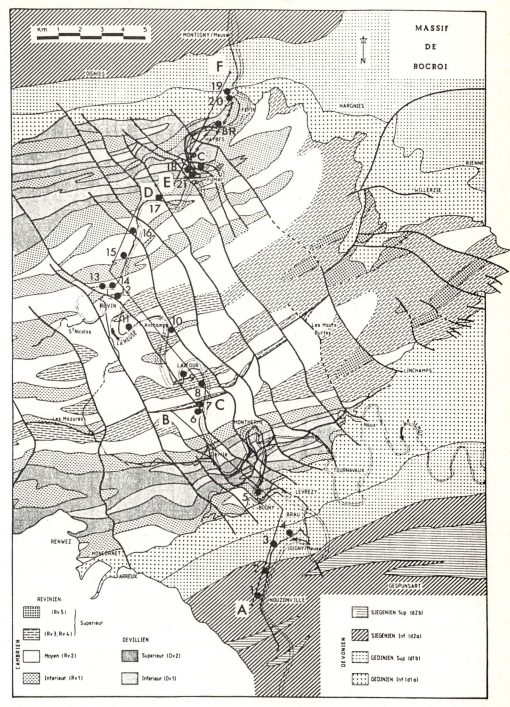

Fig. 1

A general geology map of the study region (simplified after Beugnies, 1963). The drilled sampling localities are indicated, sites 1–21, and the hand-sampled quarries Belle Rose and la Providence Charnière are labelled BR and PC.

type in the Devonian series. The megastructures of the Rocroi Massif are from north to south the Fumay anticline, the Revin syncline and the Deville-Montherme anticline. The southern portion of the Rocroi Massif is affected by the greenschist grade of metamorphism. The increase in intensity of metamorphism from north to south is expressed by the transformation of the haematite in the Devillian schists into illmenite and magnetite as well as by the presence of chloritoid crystals in Revinian and Devonian rock materials. This metamorphism is of Hercynian age. No metamorphism of Caledonian age affected the Cambrian sediments of the Rocroi Massif (Corin, 1928). A recent detailed structural study of the Rocroi Massif (Hugon and Le Corre, 1979; Hugon, 1982; 1983) demonstrates that the slaty-cleavage development affecting the Cambrian series was coeval with the peak of the Hercynian metamorphism. The folding of the Cambrian series was coeval with the slaty cleavage development and is therefore of Hercynian age. All the structures observable in the Cambrian series can be explained by a progressive ductile shearing of low-amplitude and large-wavelength NS-undulations during a northward overthrusting of Hercynian age.

Sampling

The sampling for this study was carried out along the dog-leg profile from Nouzonville in the south (point A, Fig. 1) to the north of Fepin in the north (point F, Fig. 1). The samples collected were all slates for the purpose of determining mica fabrics by means of texture goniometry. In total, 21 sites were drilled (1—21, Fig. 1) in situ (Tarling, 1971) for the magnetic fabric study. On average, 6 oriented cores of 2.5 cm diameter and about 8 cm length were taken at each site. In the quarries Belle Rose Haybes (BR, Fig. 1) and la Providence Charnière (PC, Fig. 1), loose hand-samples were taken containing reduction spots. Thirteen hand-samples were taken (B1—B7 and C1—C6, Tables 1, 2 and 3) and from each hand-sample 6 to 8 cores were drilled for the magnetic fabric method, and the remaining rock was used for the reduction spot and mica fabric studies.

Fabric Methods and Results

The Magnetic Fabric

In low fields (i.e. less than 100 Oersteds), the intrinsic magnetic susceptibility of a material (χ, Gauss Oe^{-1} cm^{-3}) is defined as $\chi = J/H$, where J (Gauss) is the magnetization produced in the material in an applied field H (Oe). The susceptibility, χ, is a second rank tensor that depends only on the intrinsic properties of the magnetic material. In high susceptibility cubic minerals such as magnetite, which was found to be the magnetic mineral responsible for the fabric in the Ardennes slates, even small strains cause major changes in the demagnetizing factors. The vectorial sum of the magnitudes of the magnetic mineral susceptibilities along three principal axes within a rock represents the rock magnetic fabric. The low field MSA can be measured using astatic magnetometers, A.C. bridges, torque magnetometers, translation inductometers, spinner magnetometers or cryogenic magnetometers (Collinson et al., 1967, Scriba and Heller, 1978).

The magnetic fabrics for this study were measured on a modified spinner magnetometer (Molyneux, 1971), known as the Complete Result Anisotropy Delineator, CRAD (Rathore, 1975). These measurements on the 200 samples were carried out in Newcastle-upon-Tyne, U.K. For reference, it should be noted that the original calibration of the Digico CRAD had been incorrect (Veitch et al., 1983 and Hrouda et al., 1983). However the data presented here have been corrected for this error. The data obtained are in the form of three principal magnetic axes $\chi_{maximum}$, $\chi_{intermediate}$ and $\chi_{minimum}$, and their orientations with respect to a fiducial mark on the sample. The orientation data are corrected for geographic north and horizontal with respect to the drilling orientation in the sampling locality. The magnitudes of the three principal axes and their orientations with respect to north and horizontal constitute the magnetic fabric of the sample. In magnetic fabric analysis, the magnitude and nature of the susceptibility ellipsoid can be expressed in terms of the axial ratios P_1, P_2 and P_3 (Stacey, 1960) defined as

$$P_1 = \frac{\chi_{max}}{\chi_{int}} ; \quad \text{lineation factor (L)}$$

$$P_2 = \frac{\chi_{max}}{\chi_{min}} ; \quad \text{anisotropy factor}$$

$$P_3 = \frac{\chi_{int}}{\chi_{min}} ; \quad \text{foliation factor (F).}$$

Table 1: Magnetic susceptibility anisotropy parameters

Site	Mean P_1	Mean P_2	Mean P_3	Mean K_a	Mean R_a	χ_{max} decl.	incl.	χ_{min} decl.	incl.
AR1	1.028	1.177	1.145	0.193	1.17	325	−59	165	−30
AR2	1.037	1.188	1.146	0.259	1.18	270	−41	124	−44
AR3	1.036	1.134	1.094	0.383	1.13	16	−61	159	−24
AR4	1.029	1.080	1.049	0.592	1.08	180	54	161	−35
AR5	1.062	1.396	1.315	0.157	1.38	186	34	211	−53
AR6	1.051	1.277	1.214	0.238	1.27	204	25	149	−51
AR7	1.080	1.215	1.125	0.650	1.21	303	−40	163	−42
AR8	1.026	1.236	1.205	0.127	1.23	306	−37	166	−43
AR9	1.042	1.157	1.110	0.382	1.15	44	−21	152	−39
AR10	1.030	1.112	1.080	0.375	1.11	304	−27	177	−51
AR11	1.039	1.115	1.073	0.534	1.11	132	25	184	−52
AR12	1.087	1.207	1.107	0.813	1.19	333	−34	173	−56
AR13	1.034	1.163	1.124	0.274	1.16	322	−34	147	−58
AR14	1.031	1.079	1.046	0.674	1.08	74	−20	181	−47
AR15	1.069	1.145	1.071	0.972	1.14	322	−58	164	−30
AR16	1.029	1.251	1.217	0.134	1.28	331	−50	156	−40
AR17	1.018	1.119	1.099	0.182	1.12	308	−49	203	−12
AR18	1.056	1.239	1.173	0.234	1.23	324	−30	165	−59
AR19	1.010	1.034	1.024	0.417	1.03	248	4	302	−83
AR20	1.008	1.061	1.052	0.154	1.06	341	6	10	−82
AR21	1.072	1.152	1.075	0.960	1.15	339	−44	194	−40
B1	1.026	1.120	1.092	0.283	1.12	264	− 3	280	−87
B2	1.020	1.141	1.119	0.168	1.14	246	− 4	87	−86
B3	1.018	1.144	1.124	0.145	1.14	124	2	83	−87
B4	1.022	1.146	1.121	0.182	1.14	93	− 4	267	−84
B6	1.022	1.115	1.091	0.242	1.11	342	1	302	−86
B7	1.033	1.137	1.101	0.327	1.13	102	3	139	−86
C1	1.035	1.189	1.149	0.235	1.18	6	0	292	−89
C2	1.025	1.195	1.166	0.151	1.19	2	2	285	−81
C3	1.019	1.195	1.173	0.110	1.19	347	0	108	−89
C4	1.023	1.214	1.187	0.123	1.21	8	2	66	−85
C5	1.028	1.196	1.163	0.172	1.19	14	− 1	269	−87
C6	1.025	1.190	1.161	0.155	1.19	283	1	196	−78

The site mean anisotropy parameters (P_1, P_2 and P_3) and the orientations of the χ_{max} and χ_{min} axes (corresponding to the strain fabric lineation and cleavage pole directions, respectively) are presented in Table 1. The ratio P_3/P_1 = E, the ellipticity of the susceptibility ellipsoid; thus if E > 1, the ellipsoid is oblate and the foliation is more developed than the lineation, and conversely if E < 1, the ellipsoid is prolate and the lineation is more developed than the foliation. In strain fabric analyses, the form parameter K_f and the intensity parameter R_f are used. The corresponding parameters for the magnetic ellipsoids are also presented in Table 1. It should be noted that the P_2 and the R_a parameters have almost the same values.

The Mica Fabric

The deformation mechanism in slates is very complex (Goguel 1967, Hobbs et al., 1973, Siddans 1976, Gratier 1978, LeCorre 1978), dependent upon the mineralogy and the strain history of the rock. The mechanism of recrystallization of micas under pressure, and reorientation due to their inherent planar geometry gives rise to a mica fabric ellipsoid. This mica fabric ellipsoid is a sub-fabric of the whole rock and can be used to give a qualitative estimate of the total strain variations. The mica fabric measurements for this study were carried out on a texture goniometer, in the reflection mode, at the University of Rennes, France. Forty-nine discs of 25-mm diameter and 4-mm thickness, cut parallel to the cleavage plane, are used. A cone of half apex angle = 70° can be covered by the goniometer. In this particular case, the degree of preferential orientations of micas is always sufficient; thus there is no need to do either extrapolation calculations or study complementary discs perpendicular to the cleavage plane. The type of correction for defocusing and absorption is an empirical one based on measurements carried out on equivalent isotropic samples. The data are in the form of mica pole density distributions on a hemisphere. In the case of orthorombic distributions, it is possible to calculate the eigenvalues a_1, a_2 and a_3 of the Scheidegger orientation tensor (1965) of the 001 mica poles. In our case, the "weighted orientation tensor" defined by Cobbold and Gapais (1979) is used. This tensor has the property of being identical to the strain tensor if the lines are possibly deformed in a homogeneous deformation. The values of X = $1/a_1$ (maximum), Y = $1/a_2$ (intermediate), Z = $1/a_3$ (minimum) mica ellipsoid axes can be determined from a_1, a_2 and a_3. The shape parameter, K, and the intensity parameter, R, are calculated, where

$$K_f = \frac{X/Y - 1}{Y/Z - 1}$$

and

$$R_f = X/Y + Y/Z - 1$$

The mica fabric data in terms of X/Y, X/Z and Y/Z together with K and R parameters are given in Table 2. The parameter K_f varies from 0 to ∞. For plane strain K_f = 1. Hence for values less than 1 the ellipsoid is oblate and greater than 1 prolate.

The Reduction Spot Measurements

Finite strain data is difficult to obtain due to the lack of suitable and reliable strain markers. Methods involving measurements of geometrically distorted regular shapes, due to straining mechanisms, give the most reliable values of finite strain. However, such strain markers are scarce. In the two quarry locations at Belle Rose Haybes (BR) and la Providence (PC) (Fig. 1), there are abundant reduction spots. We were able to select good, large hand-samples from which it was possible to get good strain measurements on the actual reduction spots, take discs for the goniometer for mica fabric strain determinations, and take drill cores for magnetic fabrik determinations. The method used for determining the finite strains on reduction spots is that of Ramsay (1967). The results are given in Table 3.

Table 2: Mica fabric strain parameters

Site	X/Y	X/Z	Y/Z	K_f	R_f
AR1	1.3	9.5	7.3	0.05	7.6
AR2	1.5	7.4	4.9	0.14	5.4
AR3	1.6	5.8	3.6	0.22	4.1
AR4	1.5	4.2	2.8	0.27	3.3
AR5	1.9	11.2	5.9	0.28	6.8
AR6	1.1	8.5	7.7	0.02	7.8
AR7	1.3	8.7	6.7	0.05	6.9
AR8	1.5	9.5	6.3	0.10	6.8
AR9	1.2	10.6	8.8	0.03	9.0
AR10	1.7	4.4	2.6	0.44	3.3
AR11	1.6	5.6	3.5	0.24	4.1
AR12	1.6	8.8	5.5	0.24	6.1
AR13	1.2	3.7	3.1	0.10	3.3
AR15	1.3	9.4	7.1	0.05	7.4
AR16	1.4	9.3	6.9	0.08	7.3
AR17	1.4	6.7	4.8	0.11	5.2
AR18	1.5	5.3	3.5	0.21	4.0
AR21	2.5	7.0	2.8	0.82	4.3
B2	1.6	10.7	6.6	0.11	7.2
B5	1.5	8.8	6.0	0.10	6.4
B6	1.8	8.1	4.4	0.24	5.2
B7	2.6	11.4	4.5	0.46	6.1
C1	1.7	9.3	5.5	0.16	6.2
C2	1.7	10.5	6.2	0.14	6.9
C3	1.7	14.0	8.3	0.05	9.0
C5	1.6	14.4	9.2	0.07	9.8

Table 3: Reduction spot strain parameters

Site	X/Y	X/Z	Y/Z	K_f	R_f
B2	1.6	10.6	6.7	0.10	7.30
B5	1.6	8.9	5.7	0.12	6.26
B6	1.9	7.9	4.2	0.27	5.08
C1	1.6	9.1	5.8	0.12	6.34
C2	1.7	10.5	6.1	0.14	6.80
C3	1.8	14.0	7.7	0.12	8.55
C5	1.9	12.9	6.9	0.14	7.75

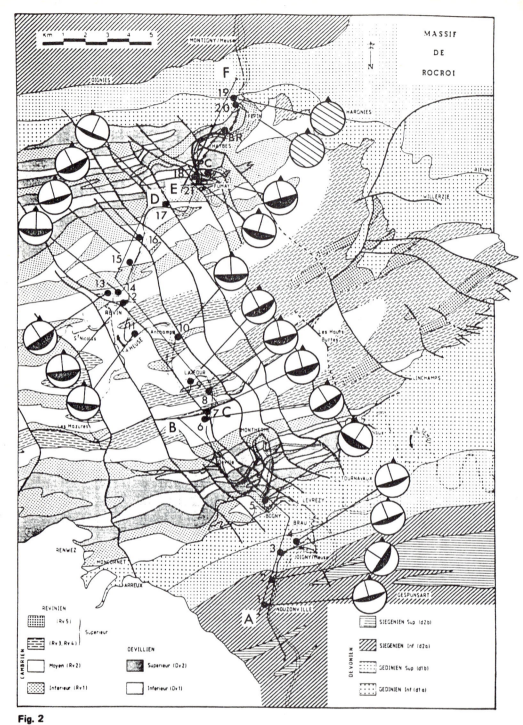

Fig. 2

The magnetic fabric orientation data is shown on lower-hemisphere equal-angle stereographic projections. The magnetic foliation planes are indicated and the corresponding declinations of the mean minimum susceptibility axes are plotted. Sites 19 and 20 have very flat-lying foliation planes and the stereograms are hatched.

Fabric and Structural Trends

Magnetic, Mica and Strain Fabrics Across the Rocroi Massif

The field-corrected orientations of the three principal magnetic axes are represented in lower hemisphere equal-angle stereographic projections (Fig. 2). The magnetic foliation planes containing the $\chi_{maximum}$ and the $\chi_{intermediate}$ axes are presented, together with the trends of the magnetic cleavage poles, the $\chi_{minimum}$ axes. An overview of the study region suggests a uniform magnetic fabric with the magnetic foliations striking approximately ENE-WSW and dipping to the SSE. This fabric trend is in very good agreement with the structural trends within the Rocroi Massif (Fig. 2). Towards the northern end of the study profile, the foliations become more flat-lying (site 18) and in sites 19 and 20 the foliations are sub-horizontal. The statistical analyses on the magnetic orientations data showed that the semi-angle of the cone of 95% confidence (α_{95}) about the mean minimum axis direction for each of the sites (containing 6 to 8 cores) was less than 10° for 17 of the sites, and not greater than 15° for the remaining sites. A rose diagram of all the magnetic cleavage poles (Fig. 3) shows a very dominant grouping of the axes along a N20°W-S20°E direction. The two other directional groupings (approximately NNE-SSW and NW-SE) are so small in comparison to the main direction that it can be concluded there is only one magnetic fabric trend across the Rocroi Massif.

A comparison is made with the orientational data obtained from the mica and reduction spot fabrics (Fig. 4). The mica fabric principal axes orientations (Fig. 4a) can be compared with the magnetic fabric principal axes (Fig. 4b) and there appears to be very good agreement in the total groupings. On the individual points level, the comparisons in a few show large angular discrepancies; however, it is pointed out that the mica fabric data are from single discs from each site whereas the magnetic data are mean values of 6 to 8 cores. If statistically significant mean orientation values from a number of discs in each site were available, then the individual site comparisons would be better and more significant. The field measured directions of the stretching lineations are plotted (Fig. 4c). Note

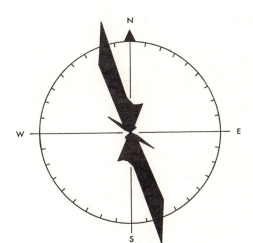

Fig. 3

A rose diagram of all the magnetic minimum susceptibility axes shows a single dominant axial grouping direction along N20°W-S20°E.

the very good agreement between the mean stretching lineation direction and the directions of the maximum fabric axes for the micas and magnetic susceptibilities.

Similarly the declinations for the mean magnetic maximum axes are compared with those of the strain ellipsoid maximum axes obtained from the reduction spot analyses. The declinations only are compared since the hand-samples taken were loose-lying and arbitrary strike directions were marked on their cleavage planes which contained the reduction spot maximum and intermediate strain axes. The drill cores for the magnetic

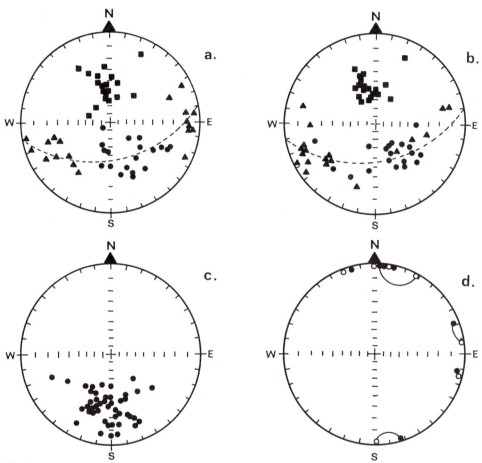

Fig. 4

A comparison of fabric orientation directions.

a) Principal axes of the mica fabric ellipsoids.
b) Principal axes of the magnetic fabric ellipsoids.
c) Field measured stretching lineation directions.
d) A comparison of orientation directions for the magnetic (●) $X_{maximum}$ and the reduction spot (○) maximum extension directions.

In Figs. a, b and c (●) = the maximum axis, (▲) = the intermediate axis and (■) = the minimum principal axis.

measurements were taken perpendicular to the cleavage planes and thus the magnetic foliation planes were found to be horizontal. It can be seen in Figure 4d that the trends of the two fabric maximum axes are very similar.

In conclusion, therefore, it can be said that the magnetic and the mica fabric ellipsoids are co-axial with one-another and that their orientations correspond to the strain axes directions as observed in the reduction spots.

Form and Magnitude Correlations of the Fabric and Strain Ellipsoids

The Ellipsoidal Forms

The shape of the fabric ellipsoids can be represented graphically using ellipticity plots (Figs. 5a and b). Such a plot is of the lineation factor (L) against the foliation factor (F), or the appropriate axial ratios. The line of slope = 1 separates the field of dominant prolateness from that of dominant oblateness.

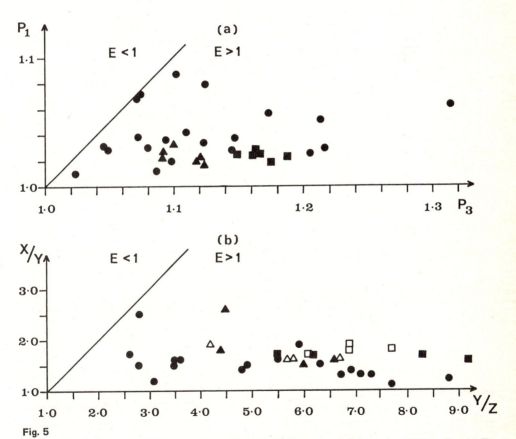

Fig. 5

a) A magnetic anisotropy plot of L vs. F shows the oblate nature of the susceptibility ellipsoid. The symbols: (●) = drilled sites; (▲) = BR sites; (■) = PC sites.

b) A corresponding strain plot of L vs. F for the mica (filled symbols) and reduction spot sites (open symbols) shows the oblate nature of the mica and strain ellipsoids. The symbols: (●) = drilled sites; (△, ▲) = BR sites; (□, ■) = PC sites.

All the fabric ellipsoids determined in this study have oblate forms with ellipticities greater than 1, (Figs. 5a and b). The degree of oblateness varies from slightly oblate to strongly oblate with the majority of the ellipsoids having well-developed oblate (compressional) ellipsoids. Note also that the K parameters given in Tables 1, 2 and 3 are all less than 1.

Although there appears to be this good agreement between the fabric forms, it should be noted that the degree of fabric development is higher in the micas and the strained reduction spots. This is apparent in that the points in the micas and reduction spot plot (Fig. 5b) lie closer to and further out along the P_3 axis. Numerically, too, the K parameters in Tables 2 and 3 are generally much lower than those in Table 1. Conversely, the R_f parameters in Tables 2 and 3 are much higher than the corresponding R_a parameters in Table 1. In conclusion, it appears that the fabric ellipsoids for all three methods have the same compressional oblate forms. However, the intensity of the fabric developments in the mica and the reduction spot fabrics is much higher than in the magnetic fabric. Such relationships between fabrics have been observed before where fabric ellipsoidal magnitudes have been compared (Wood et al. 1976, Rathore 1979, 1980, Rathore and Henry, 1982 and Rathore et al. 1983).

Magnetic Fabric to Mica and Strain Fabric Correlations

The main aim of this study was to try and establish whether or not the fabric parameters for the three fabrics from the Ardennes slates could be empirically correlated as in the case of the other studies named above. In those studies encompassing localities in Wales, the English Lake District, Scotland and Galicia, Spain, the empirical relationships found have forms of the type

$$\left(\frac{\chi_i}{\chi_j}\right) = \left(\frac{\ell_i}{\ell_j}\right)^a$$

where χ_i and χ_j are any two principal magnetic ellipsoidal axes and ℓ_i and ℓ_j are the corresponding strain or mica ellipsoidal axes. The exponent a is the correlation exponent. Hence to test for such a relationship in the Ardennes, the normalised magnetic, mica and strain fabric deviators ($\log_{10}(\chi_i/\chi_j)$ and $\log_{10}(\ell_i/\ell_j)$) were calculated and plotted in Figures 6a to 6e. Figure 6e contains all the available data and gives the linear regression relationship

$$M_i = 0.003 \pm 0.004 + (0.071 \pm 0.005)\, N_i; \quad (r = 0.819).$$

Since the regression line passes through the origin, the relationship can be expressed as

$$\left(\frac{\chi_i}{\chi_j}\right) = \left(\frac{\ell_i}{\ell_j}\right)^{0.071}$$

The correlation exponent is 0.071 and the regression coefficient r = 0.819. For a perfect fit, r = 1.0; hence the correlation in this case is fairly good. The regression lines in Figures 6a to 6d represent the more detailed analyses carried out on the fabric data. Figure 6a represents the comparison between the magnetic fabrics and the reduction spot fabrics:

$$M_i = -0.007 \pm 0.005 + (0.075 \pm 0.006)\, N_i, \quad (r = 0.942).$$

Figure 6b represents the comparison between the magnetic fabrics and the mica fabrics determined in the reduction spot hand-samples:

$$M_i = -0.006 \pm 0.004 + (0.073 \pm 0.005)\, N_i, \quad (r = 0.952).$$

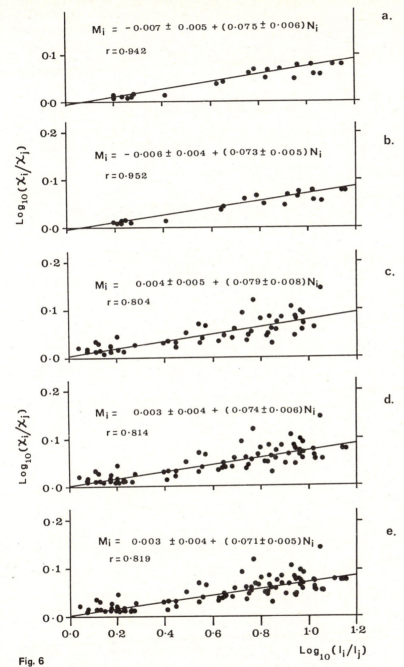

Fig. 6

Magnetic fabric to mica and reduction spot fabric correlation plots for:

a) Magnetic fabric vs. reduction spot strains.
b) Magnetic fabric vs. mica fabrics in the spot containing hand-samples.
c) Magnetic fabric vs. mica fabrics for 18 of the drilled sites.
d) Magnetic fabric vs. mica fabrics for all samples.
e) Magnetic fabric vs. mica fabrics for all sites together with the reduction spot strains.

These two regression lines imply that within the error limits of the plots, the relationship of the magnetic fabric to the reduction spot fabric is identical to that of the magnetic fabric to the mica fabrics in the same samples. This means, therefore, that the relationship between the mica and the strain fabrics is linear 1:1.

Figure 6c represents the comparison between the magnetic fabrics and the mica fabrics for the drilled sites:

$$M_i = -0.004 \pm 0.005 + (0.079 \pm 0.008) N_i, \quad (r = 0.804).$$

Figure 6d represents the comparison between the magnetic fabrics and the mica fabrics for all samples, drilled and hand-samples:

$$M_i = -0.003 \pm 0.004 + (0.074 \pm 0.006) N_i, \quad (r = 0.814).$$

Figure 6e has in addition to the mica fabric data for all samples, the reduction spot strain data from the hand samples. Since all the regression lines pass through the origin and have correlation exponents approximately 0.07 the combined regression line of Figure 6e is taken to represent the slates of the Ardennes.

Conclusions

The findings of this study are summarized in the following list of conclusions.

1. The regional magnetic fabric across the Rocroi Massif is uniform with the magnetic foliation striking approximately ENE-WSW with a steepish dip to the SSE. The magnetic poles to the magnetic foliations group very uniformly about a single dominant NNW-SSE direction.

2. The observed magnetic fabric trends are in very good agreement with the Rocroi Massif structural trends.

3. Comparisons of axial orientation data from the magnetic, mica and reduction spot fabrics show good parallelism between the corresponding principal axes for the three fabric ellipsoids. Furthermore these principal axes directions are believed to reflect the principal strain orientation in the region.

4. All fabric ellipsoids observed are oblate, suggesting a compressional straining regime.

5. The intensity of fabric development in the mica and reduction spot fabrics appears higher than that in the magnetic fabric.

6. Cross correlations of normalised fabric deviators show that the magnetic and mica fabrics as well as the magnetic and reduction spot strains are connected by the relationship:

$$\left(\frac{x_i}{x_j}\right) = \left(\frac{\varrho_i}{\varrho_j}\right)^{0.071}$$

7. This power relationship verifies the much higher degree of development of the mica and strain fabrics.

8. Furthermore, since the same correlation exponent is found for the magnetic to mica and magnetic to reduction spot correlations, it can be concluded that the relationship between the magnitudes of the mica and reduction spot fabric ellipsoids is linear and 1:1.

As a final remark, the magnitude correlation study in these Ardennes slates adds to the data in support of a possible magnitude relationship between the magnetic and strain fabrics of all rocks.

Acknowledgements

This research was conducted while Rathore held the Earl Grey Memorial Fellowship at the University of Newcastle-upon-Tyne, England. Professor Runcorn and his department are thanked for their assistance. Hugon would like to thank the Institut de Géologie at the University of Rennes and the C.N.R.S., France for their support.

References

Beugnies, A. (1963): Le Massif cambrien de Rocroi. Bull. Serv. Carte Géol. France, 270, LIX: 355–521.

Cobbold, P. R. and Gapis, D. (1979): Specification of fabric shapes using an eigenvalue method: discussion. Geol. Soc. Am. Bull., 90, 310–312.

Collinson, D. W., Creer, K. M. and Runcorn, S. K. (1967): Methods in Palaeomagnetism, Elsevier Publ. Co., Amsterdam.

Corin, F. (1928): Sur le métamorphisme d'un poudingue gedinnien entre Baneaux et Malempré. Ann. Soc. Géol. Belg., LI, 3: 100–103.

Fuller, M. D. (1960): Anisotropy of susceptibility and N.R.M. of certain Welsh slates. Nature 186, 791–791.

Girdler, R. W. (1961): Some preliminary measurements of anisotropy of magnetic susceptibility of rocks. Geophys. J. R. Astronom. Soc., 5 (3), 197–206.

Goguel, J. (1967): L'orientation des minéraux des roches sous influence de la constrainte: minéraux monocliniques et micas. Bull. Soc. Géol. Fr., 71, XI, 481–489.

Graham, J. W. (1954): Magnetic susceptibility anisotropy, an unexploited petrofabric element. Bull. Geol. Soc. Am., 65, 1257–1258.

Gratier, J. P. (1978): Schistosité de dissolution: mise en évidence et mesure quantitative des déformations. 6^{éme} R.A.S.T., Orsay, p. 192.

Heller, F. (1973): Magnetic anisotropy of granitic rocks of the Bergell Massif (Switzerland). Earth Planet. Sci. Lett., 20, 180–188.

Henry, B. (1975): Microtectonique et anisotropie de susceptibilité magnétique du massif tonalitique des Riesenferner-Vedrette de Ries (Frontière Italo-Autrichienne). Tectonophysics, 27, 155–165.

Hobbs, B. E., Means, W. D. and Williams, P. F. (1973): Folding and microfabric development in experimentally deformed salt-mica specimens. EOS Trans. Am. Geophys. Union, 54, p. 457.

Hrouda, F. (1976): The origin of cleavage in the light of magnetic anisotropy investigations. Phys. Earth Planet. Interiors 13, 132–142.

Hrouda, F. (1982): Magnetic anisotropy of rocks and its application in geology and geophysics. Geophys. Surv. 5, 37–82.

Hrouda, F. and Janak, F. (1976): The change in shape of magnetic susceptibility ellipsoid during progressive metamorphism and deformation. Tectonophysics, 34, 135–148.

Hrouda, F., Chulpacova, M. and Rejl, L. (1971): The mimetic fabric of magnetite in some foliated granodiorites, as indicated by magnetic anisotropy. Earth Planet. Sci. Lett., 11, 381–384.

Hrouda, F., Janak, F. and Rejl, L. (1978): Magnetic anisotropy and ductile deformation of rocks in zones of progressive regional metamorphism, Geol. Beitr. Geophys. 87, 126–134.

Hrouda, F., Stephensen, A. and Woltär, L. (1983): On the standardisation of measurements of the anisotropy of magnetic susceptibility. PEPI. 32, 203–208.

Hugon, H. (1982): Structures et déformation du Massif de Rocroi (Ardennes). Approche géométrique, quantitative et expérimentale. Thèse 3^{éme} cycle, Rennes Février 1982, 100 pp.

Hugon, H. (1983): Structures et déformation du Massif de Rocroi (Ardennes). Bull. Soc. géol. minéral. Bretagne, C, 15, 2: 109–143.

Hugon, H. and Le Corre, Cl. (1979): Mise en évidence d'une déformation hercynienne en régime cisaillant progressif dans le Massif cambrien de Rocroi (Ardennes). C. R. Acad. Sci. Paris, 289, D: 615–618.

Janak, F. (1965): Determination of anisotropy of magnetic susceptibility of rocks, Studia geoph. geod. 9, 290–301.

Kligfield, R., Lowrie, W. and Dalziel, I. (1977): Magnetic susceptibility anisotropy as a strain indicator in the Sudbury Basin, Ontario. Tectonophysics, 40, 287–308.

Kligfield, R., Owens, W. H. and Lowrie, W. (1981): Magnetic susceptibility anisotropy, strain, and progressive deformation in Permian sediments from the Maritime Alps (France). Earth Planet. Sci. Lett., 55, 181–189.

Kneen, S. J. (1976): The relationship between the magnetic and strain fabrics of some haematite-bearing Welsh slates. Earth and Planet. Sci. Lett. 31, 413–416.

Le Corre, Cl. (1978): Approche quantitative du processus synschisteux. L'exemple du segment hercynien de Bretagne centrale. Thèse, Rennes, 381 pp.

Molyneux, L. (1971): A complete result magnetometer for measuring the remanent magnetisation of rocks. Geophys. J. R. Astronom. Soc., 27, 429–434.

Ramsay, J. G. (1967): Folding and fracturing of rocks. McGraw Hill, New York, 568 pp.

Rathore, J. S. (1975): Studies of magnetic susceptibility anisotropy in rocks. Ph. D. Thesis, Univ. of Newcastle-upon-Tyne, England, 206 pp.

Rathore, J. S. (1979): Magnetic susceptibility anisotropy in the Cambrian slate belt of North Wales and correlation with strain. Tectonophysics, 53, 83–97.

Rathore, J. S. (1980): The magnetic fabric of some slates from the Borrowdale volcanic group in the English Lake District and their correlations with strains. Tectonophysics, 67, 207–220.

Rathore, J. S. and Becke, M. (1980): Magnetic fabric analyses in the Gail Valley (Carinthia, Austria) for the determination of the sense of movements along this region of the Periadriatic Line. Tectonophysics, 69, 349–368.

Rathore, J. S. and Becke, M. (1983): Magnetic fabrics in rocks from the Möll-Drau Valley (Carinthia, Austria). Geol. Rundsch. 72, 1081–1104.

Rathore, J. S. and Heinz, H. (1979): Analyse der Bewegungen an der Umbiegung der „Periadriatischen Naht" (Insubrische Linie/Pusterer Linie) in der Umgebung von Mauls (Südtirol) Geol. Rundsch. 68 (2), 707–720.

Rathore, J. S. and Heinz, H. (1980): The application of magnetic susceptibility anisotropy analyses to the study of tectonic events on the Periadriatic Line. Mitt. Österr. geol. Ges. 71/72, 275–290.

Rathore, J. S. and Henry, B. (1982): Strain and magnetic fabric comparisons in Dalradian rocks from the South West Highlands of Scotland. J. Struct. Geol. 4, 373–384.

Rathore, J. S. and Mauritsch, H. J. (1983): Stress-induced fracturing and magnetic susceptibility anisotropy in Bleiberg-Kreuth mining region, Austria. Tectonophysics, 95, 157–171.

Rathore, J. S., Heinz, H. and Mauritsch, H. J. (1977): Erste Untersuchung der magnetischen Suzeptibilitätsanisotropie im Bereich der Gaillinie (Nassfeldpass bis Nötsch). Anz. Österr. Akad. Wiss., Math.-Naturw. Kl., 7, 90–93.

Rathore, J. S., Courrioux, G. and Choukroune, P. (1983): Study of ductile shear-zones (Galicia, Spain) using texture gonimetry and magnetic fabric methods. Tectonophysics, 98, 87–109.

Scheidegger, A. E. (1965): On the statistics of the orientation of bedding planes, grain axes and similar sedimentalogical data. U. S. Geological Survey Professional Paper 525-c, p. 166–167.

Scriba, H. and Heller, F. (1978): Measurements of anisotropy of magnetic susceptibility using inductive magnetometers. J. Geophys. 44, 341–352.

Siddans, A. W. B. (1976): Deformed rocks and their textures. Phil. Trans. R. Soc. London, A 283, 43–54.

Stacey, F. D. (1960): Magnetic anisotropy of igneous rocks. J. Geophys. Res. 65, 2429–2442.

Stone, D. B. (1963): Anisotropic magnetic susceptibility measurements on a phonolite and on a folded metamorphic rock. Geophys. J., 7, 375–390.

Tarling, D. H. (1971): Principles and applications of palaeomagnetism. Chapman and Hall, London, 164 pp.

Veitch, R. J., Hedley, I. G. and Wagner, J. J. (1983): Magnetic anisotropy delineator calibration error. Geophys. J. R. Ast. Soc. 75, 407–409.

Wood, D. S., Oertel, G., Singh, J. (now Rathore, J. S.) and Benett, H. F. (1976): Strain and anisotropy in rocks. Philos. Trans. R. Soc. London, A 283, 27–42.

Joint Tectonics in a Folded Clastic Succession of the Variscan Orogen in the Rheinisches Schiefergebirge

D. Grzegorczyk

Westfälisches Museum für Naturkunde, Sentruper Str. 285, D-4400 Münster

H. Miller

Institut für Allgemeine und Angewandte Geologie der Universität München, Luisenstr. 37, D-8000 München 2

Key Words

Structural geology
Joint tectonics
Rock deformation
Rheinisches Schiefergebirge

Abstract

The joint pattern of folded mid-Devonian sandstones was studied in the Western Sauerland (Volme valley near Lüdenscheid). The variable dip of joints parallel to the trace of regional folds (h0l joints) can be interpreted as
a) an original property (h0l joints), or
b) an originally vertical attitude of bc (h00) joints, subsequently externally rotated during the folding (rigid body rotation), or
c) an originally vertical attitude of bc (h00) joints, subsequently internally rotated by translation processes other than b) (e.g. foliation).
Case (b) is the most frequent one, but case (c) also occurs. h0l joints formed at two different times during the folding. Early formed joints were externally and internally rotated and translated; a second set of joints formed, when about 80 % of the folding was completed; they were only externally rotated.
Vertical shear joints (hk0) formed as paired sets acute-angled on both sides of the axis of maximum principal horizontal stress, and acute-angled on both sides of the axis of intermediate principal horizontal stress.

Introduction

Mode and age of joint pattern development related to folding of rocks were often discussed. Previous papers comprise: Behrens et al. (1970) claimed a near-prefolding age of fractures within the Bavarian Folded Molasse; Maier & Mäkel (1982) tried to demonstrate a post-folding age of the joint pattern in strata of the Belgian Ardennes, whereas Bankwitz (1965, 1966) thought of a synchronous jointing process during the folding.

Joints are found in unfolded sediments as well as in polyphasely folded and metamorphosed rocks. Hence, it seems to be probable that joints are forming repeatedly during the history of a rock, and that the present-day joint pattern of a folded series is the result of several processes.

The joint pattern was investigated by us on the outcrops of Devonian beds along the Volme valley in the Sauerland area (Rheinisches Schiefergebirge) E of the Rhine river (Fig. 1). The area of research is situated on the northern limb of the Lüdenscheid syncline. The rocks are composed of sandstones and slates, rarely with generally low content of CaCO$_3$.

Most outcrops belong to the Mühlenberg, Brandenberg and Lower Honseler Formations of Late Eifelian to Early Givetian age. Measurements of the joints orientation were made on sandstones only.

In the northern Rheinisches Schiefergebirge, folding produced large regional synclinoria and anticlinoria which can be related to synsedimentary basins and rises (Oncken 1982,

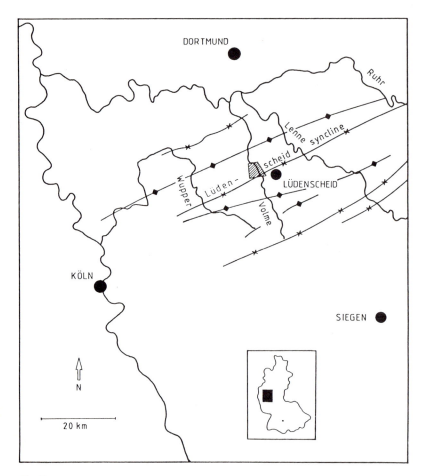

Fig. 1
Area of investigation.

1984). Minor folds show wave lengths of meters to tens or hundreds of meters. Due to the generally poor exposure, it is often impossible to define the exact position of fold axes in the field. The axial planes are generally vertical, sometimes with a steep SSE dip. The fold axes are oriented WSW—ENE as is common in this part of the Variscan mountain chain.

Geometry and Modes of Origin of Joints in Folded Rocks

The geometry of joint sets has been described, for example, by Sander (1948): The most common joints (Fig. 2) are those perpendicular to the fold axis (ac) and those parallel to the axial plane (bc). Various shear joints run parallel to the fabric b axis (h0l), to the a axis (0kl), and to the c axis (hk0); the latter are the most frequently mentioned shear joints.

While the originally vertical joint sets of the ac and bc positions ("orthogonal joint system"; Bock 1980) are reported also from laterally non-stressed sediments, shear joints only develop in three-axially stressed rocks.

Following the laws of mechanics (see Mohr circles), shear joints form as a conjugate set with angles less than 45° with the major principal stress (a-) axis. W. Schmidt (1932)

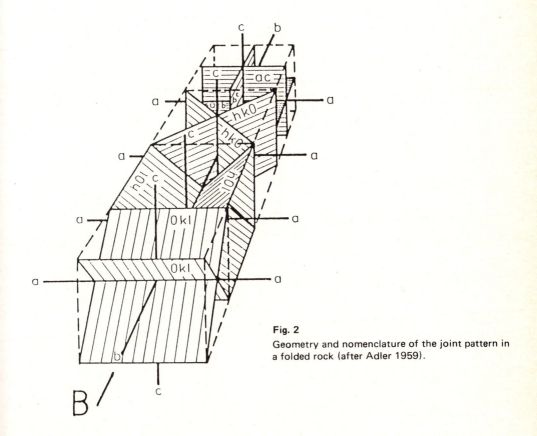

Fig. 2
Geometry and nomenclature of the joint pattern in a folded rock (after Adler 1959).

demonstrated that shear fractures may also develop at an acute-angle on both sides of the intermediate principal stress axis, a hypothetical inference that is frequently developed in the nature, but rarely mentioned. h0l-joints, i.e. joints parallel to the b-axis, dipping to both sides and with variable angles, are commonly found. They should enclose angles less than $45°$ with the a-axis, i.e. they should have the positions of low angle faults. Such faults and fractures occur, but they are by far not as frequent as steeply dipping h0l-joints, the appearance of which cannot be explained by current mechanical concepts. This made us suppose that h0l-joints in folded rocks commonly are bc-joints, rotated from a vertical or steeply inclined position into their present position by folding (rigid-body or external rotation).

On the other hand, steeply or vertically dipping joints parallel to bc may also have formed by rigid-body translations or internal rotation (foliation processes) of formerly inclined joints during folding.

Summing up, there are three modes of formation of b-parallel joints:

a) originally non-vertical formation,

b) originally vertical formation of bc-joints, subsequently rotated externally during folding,

c) originally vertical or non-vertical formation of bc- or h0l-joints, subsequently rotated internally by translation (foliation processes).

We shall show that case b) is a frequent one, but that case c) is also important in fold belts of moderate deformation rates like the Rheinisches Schiefergebirge.

Age and Mode of Origin of B-Parallel Joints in the Area of Investigation

Method

If a joint formed in bc position after folding, it shows a vertical dip. If it formed before folding, a reconstruction of the joint set together with the bedding plane to the latter's original horizontal position must result in a pre-folding vertical position of the h0l- (former bc-) joint.

Fig. 3 schematically shows the path of fracture poles of different symmetry and different time of formation within an equal-area (Schmidt) net during folding of a bedding plane (ss). Obviously ac-joints are apparently not rotated due to their position perpendicular to the rotation axis.

If the bc plane formed during the folding process, there will be a stage during the reconstruction of the bedding plane, where the h0l plane occupies a vertical position. This stage of the bedding plane (its dip at the time when the bc plane formed) is called "time of joint formation". In a diagram "dip of bedding at the time of joint formation" versus "present dip of joint" (Fig. 4), joints which formed after folding fall on a straight line inclined $45°$; joints formed before folding are found on the abscissa; joints formed at any stage during the folding get their position between these lines. The gradient of the lines on which joints concentrate indicates the relative age of joint formation.

It is important to state that instead of "vertical position" we really have to say "position parallel to the deformation (bc) plane", if this plane is not exactly perpendicular to the earth's surface. In our reconstructions we took care to consider that, controlling the dip of the axial plane.

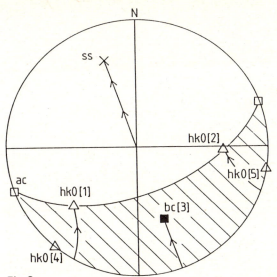

Fig. 3

Paths of poles of vertically formed joints of variables age in an equal area net during folding of the bedding plane (ss). hk0 (1) and hk0 (2) are the oldest joints, bc (3) formed after one third of folding, hk0 (4) and hk0 (5) formed after folding. ac shows no apparent displacement.

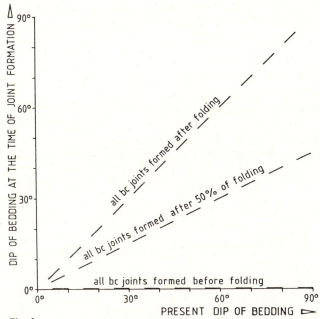

Fig. 4

Schematic diagram showing the relations between the folding process and the relative age of joints.

bc-Joint Patterns From the Volme Valley

Fig. 5 (a–d) shows some typical fracture patterns of outcrops of the area of investigation. It is evident that not only the h0l joints are partly rotated together with the bedding planes by folding, but also the hk0 joints often lie on the same great circle opposite to the pole of the bedding plane. As hk0 joints hardly can have formed in a position other than parallel to the c axis, this is evidence for a c-parallel origin of the present h0l joints, i. e. for their origin as bc-joints.

Fig. 6 shows a diagram of reconstructed h0l-joints as defined in Fig. 3 and 4. Each sign stands for the reconstructed mean value of h0l-joints (former bc-joints) of an outcrop (30 outcrops altogether). Obviously the reconstructed positions lie on two straight lines, one almost crossing the origin of coordinates (line A), the other one crossing the abscissa at about $30°$ of "present bedding plane" (B). As shown in Fig. 4, the first mentioned joints can be interpreted as having formed at a very late folding stage of the bedding planes; to be exact, after having completed about 80 % of the total folding of the respective area.

Theoretically, an extremely strong internal rotation of early formed by-parallel fractures by foliation processes would produce the same aspect of now nearly vertical planes. However, there is evidence that in the investigated area internal rotation (foliation) has not been very strong (Schreiner 1982). Hence it follows that line A very probably is not influenced by internal rotation.

The second straight line (B) which runs nearly parallel to A intersects the abscissa at about $30°$ of "present dip". As explained above, all fractures having formed before folding finished are located within a field between the $45°$ line through the origin and the abscissa. So it is clear that these fractures formed at an early stage of folding. Such joints, however, should lie on a straight line which runs through zero as do all externally rotated bc-joints.

Similarly several other diagrams from areas of the Volme valley and its vicinity show the same phenomena (Grzegorczyk 1983): h0l-joints concentrate near a $45°$ inclined straight line and along another one which intersects the abscissa at $30°$; sometimes joints scatter in the field between these two lines. This distribution of joint positions as well as the $30°$ value of the abscissa intersection must have a meaning:

We suggest a virtual rotation of the joint positions around a rotation axis that does not intersect the origin of coordinates. Thus the second set (B) of joints cannot be interpreted as having formed early and subsequently externally rotated. These joints must rather have been partly deformed by an internal rotation of early formed joints caused by foliation processes. The distance a in Fig. 6 may be considered as a measure for the extent of internal deformation of the rock during folding and foliation. The counter-clockwise rotation of the joint line B indicates an increase of internal deformation along with the intensity of folding. Schreiner (1982) has demonstrated that in the investigated area an exact determination of the overall strain rates is very difficult. He quotes a strain rate of 26 % for the Lüdenscheid syncline, but shows that the locally observed strains vary strongly.

Two Sets of Paired hk0-Joints as a Usual Pattern in Folded Rocks

Shear fractures parallel to the c-axis of the fabric are very common features in folded rocks. According to the commonly considered laws of mechanics, a pair of c-parallel shear fractures is situated on both sides of the a-axis (maximum horizontal normal stress) including an angle of less than $90°$ between each other. However, there is the possibility that another pair of shear fractures may form at both sides of the intermediate

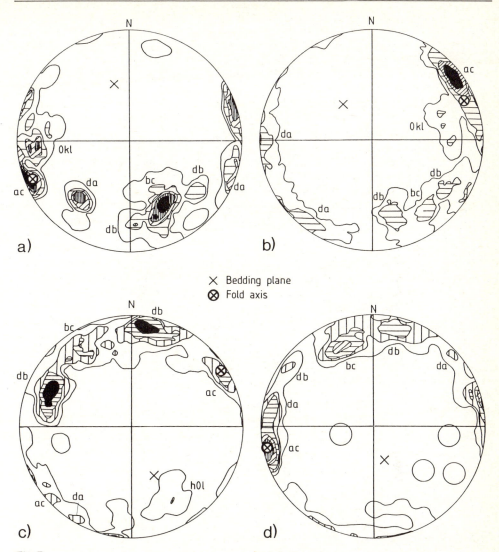

Fig. 5

Four examples of joint patterns. ac, bc, 0kl, h0l, d_a, d_b are genetically classified joints (see Fig. 2). d_a are paired hk0-joints striking acute-angled to the a-axis; d_b are paired hk0-joints striking acute-angled to the b-axis.

a) Mintenbecke valley, road cutting N Tinghausen; r 34 01 950/h 56 74 360; sandstone of the Lower Honseler Schichten, 50 joints, ss 165/44, B 248/07. — The diagram shows early formed ac, bc and hk0 joints in a S dipping limb.

b) Halver valley, road cutting N Halver; r 33 97 670/h 56 77 380; sandstone of the Brandenberg-Schichten, 50 joints, ss 142/34, B 67/10. — Joint sets in a S dipping limb. All joints are of intermediate age; hk0 joints acute-angled to b (d_b) and a (d_a) are both well developed.

c) Volme valley, road cutting E of Stephansohl; r 33 99 800/h 56 77 510; sandstone of the Lower Honseler Schichten, 100 joints, ss 336/40, B 50/06. — An example for relatively young joints, which are nearly vertical, while the bedding plane is dipping with an angle of about 40°.

d) Volme valley in Schalksmühle, road cutting NW of the cinema; r 33 97 350/h 56 79 795; sandstone of the Mühlenberg-Schichten, 100 joints, ss 343/26, B 260/03. — An example for a NNW dipping bedding plane, showing the excellent conformity of the B-axis with the poles of ac-joints, and two pairs of hk0-joints.

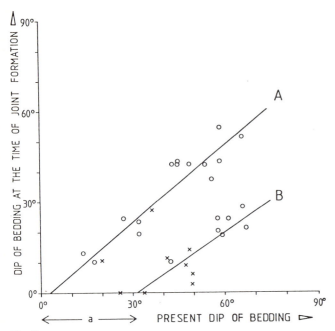

Fig. 6

Former bc-joints of 30 outcrops related to the dip of the bedding
planes. Circles for joints related to S dipping bedding planes; crosses
for joints related to N dipping bedding planes.

stress axis. Cursory joint analyses frequently do not reveal this secondary set of fractures.
hk0-joints at both sides of the b-axis (d_b in the diagrams) are very common in the investigated area (Fig. 5). They partly formed during, partly after the folding when the
maximum horizontal stress decreased, while the stress along b remained constant. Their
geometric distribution strongly varies, so that an exact age classification is difficult.

Conclusions

In the northern Rheinisches Schiefergebirge, b-parallel joints formed at two well-defined
stages of folding. An older set was subsequently deformed by internal and external rotation due to the folding of the strata involved and due to foliation processes. A younger
set formed, when the bedding planes achieved about 80 % of their final inclination. They
were externally rotated during the last increment of folding.
The b-parallel joint sets in h0l positions generally are folded former bc-joints. The existence of a second c-parallel system of hk0-joints, which include angles less than 45°
with the b-axis, in addition to the well-known system of shear joints including less than
45° with the a-axis, has been rarely acknowledged so far. Joints of this kind are widespread in many mountain chains, but poorly known because of the scarcity of systematic
joint fracture research.
All these phenomena can only be identified statistically; an individual analysis of only a
few particular outcrops is not appropriate for detailed work on joint tectonics.

References

Adler, R.E. (1959): Eine neue Methode. Erfassung und genetische Deutung von Schlechten am Kohlenstreb. Bergbau-Rdsch., 12, 469—480.

Bankwitz, P. (1965): Über Klüfte I. Beobachtungen im Thüringischen Schiefergebirge. Geologie, 14, 241—253.

Bankwitz, P. (1966): Über Klüfte II. Die Bildung der Kluftfläche und eine Systematik ihrer Strukturen. Geologie, 15, 896—941.

Behrens, M., H. Frank, K. Höllein, W. von Spaeth and P. Wurster (1970): Geologische Untersuchungen im Ostteil der Murnauer Mulde. Z. dt. geol. Ges., 121, 197—224.

Bock, H. (1980): Das fundamentale Kluftsystem. Z. dt. geol. Ges., 131, 627— 650

Grzegorczyk, D. (1983): Klufttektonik und Verformungsablauf der variszischen Gebirgsbildung im West-Sauerland (Bereich Volme-Tal). Diss. Univ. Münster, 311 pp.

Maier, G. and G.H. Mäkel (1982): The geometry of the joint pattern and its relation with fold structures in the Aywaille area (Ardennes, Belgium). Geol. Rdsch., 71, 603—616.

Oncken, O. (1982): Determinierung und Entwicklung großtektonischer Strukturen im nördlichen Rhenoherzynikum (Beispiel Ebbe-Antiklinorium). Diss. Univ. Köln, 189 pp.

Oncken, O. (1984): Zusammenhänge in der Strukturgenese des Rheinischen Schiefergebirges. Geol. Rdsch., 73, 619—649.

Sander, B. (1948): Einführung in die Gefügekunde geologischer Körper. I. Allgemeine Gefügekunde und Arbeiten im Bereich Handstück bis Profil. 215 pp., Springer, Wien.

Schmidt, W. (1932): Tektonik und Verformungslehre. 208 pp., Bornträger, Berlin.

Schreiner, M. (1982): Tektonische Verformungsanalyse im Remscheid-Altenaer Sattel, in der Lüdenscheider Mulde, im Ebbe-Sattel und in der Attendorn-Elsper Doppelmulde (östliches Rheinisches Schiefergebirge). Geotekt. Forsch. 63, 1—99.

Some Remarks on the Position of the Rhenish Massif Between the Variscides and the Caledonides

G. Hirschmann

Thieshof 20, D-3000 Hannover 51, Federal Republic of Germany

Key Words

Variscides
Internides
Externides
pre-Variscan development
Polarity
Caledonian event

Abstract

On the background of the repeated discussion on the mid-European Caledonides, their relations to the Variscides and the present plate-tectonic models a simplified synopsis of selected geological facts is presented including the Proterozoic-Palaeozoic development in a section from the Bohemian massif/Upper Rhine region through the Rhenohercynian zone to the Brabant massif. An attempt is made to demonstrate that the polarity within the Variscan zones involves subsidence, magmatism and deformation processes of the Cadomian, Caledonian and Variscan stages of development, thus showing the intimate relations between them. More or less limited pre-Carboniferous molassoid sedimentation, indicating the beginning cratonization and marking the following transition to the Variscan stage, are decreasing in age from the internides (Cambrian — Tremadocian, concluding a restricted Cadomian development) to the externides and the foreland (Old red, concluding the Caledonian development). With increasing duration and importance of the Caledonian cycle, the Variscan development is diminishing in time and intensity from south to north. The Variscan externides developed, in this view, in a rather mobile area without a coherent or widespread Cadomian or Caledonian basement. Processes in the internides synchronous to the Caledonian development in the north should be considered as belonging to the Variscan cycle.

Introduction and Present Discussions

Since Kossmat (1927), the Rhenish massif is recognized as a part of the external zones of the Mid-European Variscides. The position of the latter between Laurussia and the Caledonian Paleuropa on the one hand, and Gondwana on the other hand, has been explained by Stille (1946, 1948, 1951). The polarity of the development from the Variscan internides to their northern externides is known and generally accepted at least for parts of the orogene. Most authors thereby rely on the study of the Variscan era sensu stricto (that means the Devonian — Carboniferous time), whereas the role of the older elements (mainly in the internides) is under controversial discussion, e. g. non-regenerated massifs of Assyntian or Caledonian age (Stille, 1951); parts ("Teilblöcke") of a dissected basement (Brause, 1970); Gondwana-derived microcontinents (Ziegler 1982, 1984); presence of old (Archaic — Lower Proterozoic) crustal units analogous to those in the Baltic shield = Baltic-Bohemian craton (Wienholz, Hofmann & Mathé 1979); older parts of a succession of repeated incomplete orogenic cycles (Hirschmann, Hoth & Lorenz 1968; Zoubek, 1974).

The application of the geotectonic and geomagmatic cycle according to Stille has proven useful until recently and revealed numerous important facts and regularities in the Mid-European Variscides. New possibilities of interpretation, but also excited discussions, resulted from the introduction of plate-tectonic models to this area. Thereby the classical orogenic cycle can be interpreted as a succession of rifting and subducting or collision processes (opening and closing of the Iapetus and the Proto-Tethys between Laurentia, Fennoscandia and Gondwana — Ziegler 1982, 1984; continental rifting and subfluence — Behr 1978, Weber & Behr 1983). The localization of the rifting and subduction processes, and their extent and ages are still not very well determined. Unresolved remains the question for the crustal state during the history since the Proterozoic (oceanic or continental? — see for discussion Weber & Behr 1983; intermediate crust ("Schelfkruste") — Brause 1979a).

In the last decade, the role of Caledonian and Acadian tectonic events within the Variscides has been subjected to repeated discussion. The increasing number of age determinations giving corresponding values has been interpreted in a quite different way (e.g. Schmidt 1976, Dornsiepen 1979, Hofmann et al. 1979, Ziegler 1982, Weber & Behr 1983).

According to Ziegler (1982, 1984), the development of the Saxothuringian-Barrandian and the Cornwall-Rhenish-East Sudetic basins took place in connection with the N-dipping Proto-Tethys B-subduction zone by the alternation of back-arc-rifting and back-arc-compression within the overriding northern plate. The Variscan externides developed, in this view — similar to that of Krebs (1978) — on a basement of Caledonian consolidation. This model can explain the general succession of consolidation from N to S within Europe according to Stille, but the role of the well-known polarity within the Variscides remains uncertain. Better suited in this respect seems to be the model proposed by Weber & Behr (1983) with a south-dipping subduction zone (respectively, two symmetrical subduction zones) withdrawing stepwise from the internal parts of the orogene. A model proposed by Brause (1980, "Differentialmobilismus") shows some similarities.

As stressed by Behr (1978) et al., it is necessary to consider the whole Proterozoic and Palaeozoic history of the region in order to achieve a valid geodynamic interpretation. For the Rhenohercynian zone, this is impossible without regarding the Variscan internides, where the knowledge within the last two decades has been essentially improved. On this basis, some considerations concerning the relationship between Variscides and Caledonides and the position of the Rhenish massif in this setting shall be presented. It is, thereby, not intended to discuss the character and scope of plate movements and nappe tectonics.

Pre-Variscan Development

Fig. 1 displays a schematical and certainly incomplete synopsis of some important features of the individual zones from Proterozoic to Late Palaeozoic times. Though the stratigraphical limits between the Cadomian, Caledonian and Variscan cycles are quite unclear (Kvale 1977) and not synchronous, it shall be roughly distinguished between pre-Caledonian (till the Lower Ordovician) and Caledonian developments (till the Lower or Middle Devonian).

Pre-Caledonian Development

Moldanubian zone: The most important geosynclinal subsidence took place in the Upper Proterozoic (Brioverian Supergroup). The lower limit of this unit may be in the order of 1000 Ma (profiles in the Teplá-Barrandian-Železné hory region). The existence of an individual older complex (Moldanubian Supergroup according to Zoubek 1974 et al.)

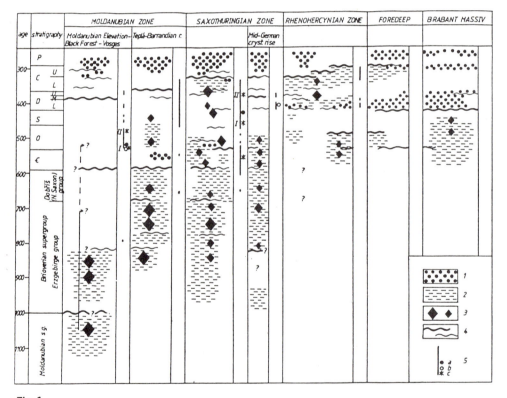

Fig. 1

Schematical synopsis of the Proterozoic-Palaeozoic development in a section from the Bohemian massif/Upper Rhine region through the Rhenohercynian zone to the Brabant massif.

Symbols: **1** molassoid sedimentation, **2** geosynclinal, mainly detritic sedimentation of greater thickness, **3** geosynclinal volcanism (maxima), **4** unconformities of various importance, **5** age determinations indicating metamorphic processes; **5a** granulitic metamorphism, **5b** intrusion of orthogneisses, **5c** anatexis (I, II — anatexis I, II in the Black Forest/Vosges and in the Saxonian granulite massif respectively).

is unresolved (for discussion see Kodym 1976). The geosynclinal basic to bimodal magmatism comprises a greater span than was earlier supposed (Holubec 1966). It has its last maximum in the Dobříš group (Röhlich 1965). Besides the rather clear Cadomian folding prior to the Middle Cambrian, older disconformities are reported mainly from the section of the Železné hory — Kutná Hora region (Kettner 1946, Urban 1972, Chaloupský 1978). They can be presumed to be in the range of 800 to 650 Ma. The youngest of them is the so-called "Eisengebirgische Phase" at the base of the Dobříš group. In general, the age determinations don't reflect these older tectonic events; only very few values are older than 600 Ma. In the Barrandian area the pre-Caledonian development is finished by the deposition of the molassoid Cambrian complexes in a rather narrow zone.

Saxothuringian zone: The thick detritic sediments of low maturity of the Proterozoic in the Erzgebirge are of a comparable age with those in the Teplá-Barrandian area (Erzgebirge group). Equivalents are present in other parts of the Saxothuringian zone. Flyschoid sediments of the Dobříš group (= North Saxonian group) are widespread and of great thickness in northern and eastern Saxony (Hoth et al. 1979, Lorenz & Burmann 1972, Hirschmann 1978). In the western part of the Fichtelgebirge-Erzgebirge anticlinal zone, the detritic sedimentation (pelitic to psammitic with varied intercalations) continues till the Early Ordovician. Geosynclinal magmatism occurs with several maxima within the Erzgebirge group and in the Cambrian. The Cadomian folding is proved in the Lausitz anticlinal zone between the North Saxonian group and the Lower Cambrian or the Tremadocian; it is supposed to exist in parts of the Fichtelgebirge-Erzgebirge anticlinal zone and is lacking in the Schwarzburg anticlinorium (Hirschmann 1966, Stettner 1974, Bankwitz, E. & P. 1975). Older disconformities with folding and partly with granitoid intrusions are documented within and after the Erzgebirge group in the Lausitz anticlinal zone and the Elbe valley zone (Hirschmann 1975, Bankwitz, P. & E. 1982, Frischbutter 1982). In the Doberlug synclinal zone, the flyschoid Middle Cambrian series is probably folded prior to the Ordovician (Sardic phase — Brause 1969). In the Vogtland area, a disconformity within the Lower Ordovician is marked by the transgression of the slates with Phycodes (Douffet 1970). Vast anatectic processes took place in the Lausitz anticlinal zone probably during the Lowermost Palaeozoic. They are documented by a first group of age values similar as in the eastern part of the Erzgebirge (Wienholz, Hofmann & Mathé 1979). In eastern and northern Saxony, the locally developed Tremadocian quartzites show similarities with molassoid deposits and thus mark the incipient crustal consolidation.

Caledonian Development

Moldanubian zone: Though there are repeated discussions on the Palaeozoic age of parts of the Moldanubian groups (Andrusov & Čorna 1976, Pacltová 1980, Tollmann 1982), an intense Cambro-Silurian subsidence is not very probable for the Moldanubian elevation. The deformation of the mainly Proterozoic rocks and their metamorphism are, according to the stratigraphic evidence in the Black Forest and the Vosges, pre-Middle Devonian. The age determinations in the range from 525 to 475 Ma suggest an (Cambrian to) Ordovician age for anatexis I, intrusion of orthogneiss-magmatites, regional metamorphism and anatexis II (Hofmann & Köhler 1973, Weber & Behr 1983). Within the Barrandian marine Palaeozoic sequence, only the Ordovician shows detritic sedimentation of greater thickness. At least two maxima of geosynclinal volcanism appeared in the Lower Ordovician and in the Silurian, but the folding is early Variscan (after Middle Devonian). Caledonian molassoid sediments are lacking.

Saxothuringian zone: From the Middle Ordovician to the Lower Devonian, a rather reduced sedimentation prevailed. A strong basaltic — keratophyric volcanism is developed in the Ordovician of the "Bavarian" facies/vicinity of the Münchberg massif. In the other synclinal zones, a maximum of geosynclinal volcanism exists near the Silurian-Devonian boundary (e. g. in the Upper Silurian of the western Sudetes — Svoboda & Chaloupský 1966). Tectonic deformations and discordances in early to late Caledonian time are known mainly from the western Sudetes, where a disconformity in the Upper Ordovician is followed by a strong, but regionally limited deformation between the Silurian and the Devonian (Chaloupský 1966). It can be still traced in the eastern part of the Lausitz region. The other synclinal zones are characterized by an uninterrupted Silurian-Devonian boundary. At least in parts of the Mid-German crystalline rise (Saar 1), the Middle Devonian transgressed on an eroded crystalline basement, the metamorphism of which is probably of Lower Devonian age. According to Weber & Behr (1983), the succession of anatexis I, granulite metamorphism and anatexis II occurred in the Saxonian "Granulitgebirge" and the southern parts of the Saxothuringian zone in the interval from 460 to 380 (−330) Ma. The relationships between these processes in the deeper crust and the development in the more superficial geosynclinal areas are still uncertain. Analogous to the Moldanubian region, an epi-Caledonian molassoid sedimentation is unknown and the delimitation of the Caledonian and Variscan developments is fluent.

Rhenohercynian zone: Because of the rare outcrops of pre-Devonian complexes, the information on the Caledonian stage is limited. In the northern Rhenish massif, early Caledonian folding (Middle Ordovician) is recorded from the (southern) Ardennes massifs, the continuation of which is supposed to be in the northern Sauerland. Further north, the folding is situated at the Silurian-Devonian boundary (Fourmarier & Michot 1964). Rather unresolved remains the role of the so-called "Eckergneis" in the Harz mountains (392 Ma). In the southern part of the Rhenohercynian zone, it is difficult to solve the problems connected with the so-called "Metamorphic zone" (Soonwald — Taunus — Lower Harz). Some authors consider a pre-Devonian metamorphism. The available age determinations are not convincing in this respect. On the other hand, in the Harz mountains the existence of Ordovician to Devonian complexes has been proven within this zone (Burmann 1973, Schwab 1977). Similarly, in the Soonwald and Taunus, the participation of Devonian rocks is supposed within the Metamorphic zone. The enclosed metamagmatites are of intermediate to acid character (Anderle & Meisl 1977). In the Rhenish massif, the prominent geosynclinal accumulation started in the Lower Devonian. Remarkable are the Old-red-influenced clastics derived from Caledonian folded regions in the north. Less pronounced are similar influences from the Mid-German crystalline rise in the south.

Subvariscan foredeep and foreland: Corresponding to the Brabant massif in the region immediately south of it, the Caledonian deformation took place between Ludlowian and Gedinnian. Of some importance is the result of the borehole Soest-Erwitte, in which the metamorphic Ordovician shales have a minimum age of 344 Ma (Krebs 1978, Ahrendt, Hunziker & Weber 1978). The often proposed hypothesis of a pre-Variscan "East Elbe massif" (Precambrian or Caledonian) is based only on the intense geophysical anomalies, which may originate in the deeper crust and/or the mantle.

North of the foredeep, the development of the Brabant massif is in good agreement with the tendencies of increasing importance of the Caledonian event and the shifting from intra-Ordovician to late Caledonian deformation in a northern direction. The assumption of a Cadomian cratonized basement as a continuation of the Midland craton underneath the Brabant massif is speculative, though the exact delimitation of the two units west of

London is still badly defined (Dunning & Watson 1977). The Old red is developed on the Caledonian Brabant massif as well as on the Midland craton, there conformably overlying the Silurian.

Polarity of the Development

In describing the polarity of the mid-European Variscides, most authors consider the N-migration of sedimentary troughs, the vergencies, the increasing younger age of the main folding and of the flysch-molasse boundary from the internides to the northern externides or within individual zones. The magmatism shows a similar polarity. The maxima of geosynclinal volcanism are shifting from the Barrandian area (Lower Ordovician, Silurian) via the Saxothuringian zone (Upper Silurian/Lower Devonian, Upper Devonian) to the Rhenohercynian zone (Middle Devonian, Lower Carboniferous). Tendencies of this kind are observable for the late or post-tectonic granitoids and subsequent volcanics (e.g. Lippolt, Schleicher & Raczek 1983, Brause 1979b). When regarding mainly the development of the Variscan troughs, it is justified to stress the differences between internal and external parts of the orogene (stationary resp. migrating troughs — Krebs 1977, "saxotype" resp. "siegenotype" — Brause 1970).

The above discussed data are also suitable for an interpretation of polarity involving pre-Variscan elements. These are to be deduced from the anticlinal areas, which are alternating with the synclinal ones in all zones of the orogene, even if to a different degree. Since there are no widespread areas of Cadomian or Caledonian cratonization, the development of the older complexes outcropping in the anticlinal areas is closely connected with the Variscan development in the synclinal regions.

The beginning of geosynclinal subsidence is still not defined, but it may be in the Middle Proterozoic. The important pre-Variscan geosynclinal subsidence lasted to the upper limit of the Proterozoic (Dobříš group) in the Moldanubian zone, at least in parts to the Cambrian or Lower Ordovician in the Saxothuringian zone, to the Middle Ordovician in the northern Rhenohercynian zone, and to the Silurian in the Brabant massif.

Since the Moldanubian tectogenesis (as an equivalent of the Dalslandian or Grenvillian) has become doubtful in the core of the Bohemian massif, the oldest disconformities within the Brioverian Soupergroup must be considered as pre- or early Cadomian deformations, having acted in parts of the Moldanubian and Saxothuringian zones. A similar distribution can be assumed for the Cadomian folding, though it is of some greater importance and extent. Apparently its influence is decreasing further to the north. Sardic discordances can be observed mainly in the Saxothuringian zone, whereas the early Caledonian folding has been active in the northern Rhenohercynian zone (Ardennes, Lippstadt anticline?). The young Caledonian (to Acadian) movements have been efficient in a few limited parts of the Saxothuringian zone, but they obtained their maximal importance only outside the Variscides (Brabant massif, southern North sea).

As concerns available age determinations, it seems that geologically corresponding metamorphic events (e.g. anatexis I, granulitic metamorphism, LP-metamorphism, anatexis II) show a temporal shifting from Cambro-Ordovician (Moldanubian zone) to Silurian-Devonian (Saxothuringian zone). Likewise of special interest are the first (pre-Variscan) symptoms of beginning or increasing crustal consolidation, documented by molassoid sediments, and their stepwise "migration" from south to north (Cambrian — Ordovician — Old red). Correspondingly there is a migration of the transition phase to the Variscan development from the Lower Ordovician to the Upper Devonian/Lower Carboniferous. A good example is given by Franke et al. (1978) from SW-England. In contrast to older molassoid deposits, the Permo-Carboniferous molasse covered not only parts of the Variscan orogene, but it is present in all zones and even in essential parts of the Caledonides.

Conclusions

1. The polarity of the mid-European phanerozoic belt includes not only the Variscan but also the pre-Variscan stages of development (subsidence, deformation, magmatism and molassoid sedimentation).
2. Besides the still not clarified Moldanubian tectogenesis[1]), the most important pre-Variscan events are the Cadomian and the Caledonian ones (in a broader sense) with several phases of deformation in each case. A sharp distinction is hardly possible. The older phases (from Uppermost Proterozoic to lowermost Ordovician) are of greater importance in the internides; the younger (Ordovician to Silurian-Devonian boundary) within the externides or north of them. The degree of partial or temporary cratonization is increasing from Cambrian (Teplá-Barrandian) to Devonian (Brabant massif).
3. Nevertheless, a widespread Cadomian basement (coherent over greater areas or dissected into individual blocks — Stille 1951, Brause 1970, Ziegler 1982 et al.) is lacking even in the internides. Its existence in the externides is problematic.
4. The duration and importance of the Variscan development decreases with increasing importance of the Caledonian events from the internides (starting from Early or Middle Ordovician) to the externides (starting from Early to Late Devonian). The metamorphic and magmatic processes in the internides during Ordovician to Devonian time (i.e. synchronous to Caledonian events north of the Variscides) therefore should be considered as belonging to the Variscan rather than to the Caledonian development.
5. Caledonides and Variscides are intimately connected with each other in space and time (Hoth, Hirschmann & Lorenz 1970). The sharp discordant delimitation between them is, according to Matthews (1978), problematic. It seems unnecessary to look for more internal and strongly metamorphic parts of the "mid-European Caledonides" underneath the Rhenish massif or other parts of the Variscan externides (e.g. Krebs 1978) or to consider the development of the Variscan externides as an overprint of a Caledonian basement (Ziegler 1984). Despite the existence of a number of Caledonian folded cores, the Variscan externides can be regarded as developed in a long-lasting mobile area. This is in good agreement with thoughts on the development of the mid-European lithosphere proposed by Bankwitz et al. (1979).
6. In agreement with Weber & Behr (1983), the development of the mid-European Variscides can be seen as a long-termed process from Proterozoic to Palaeozoic. A similar view is valid for the Caledonides at least since 1000—1100 Ma (Kvale 1977, Watson 1977). It seems to be possible that the subsidence in the region of the European phanerozoic mobile belts started already shortly after the Archaean-Early Proterozoic cratons in northern Europe and America had been finally formed during Svecokarelian time (1700—1800 Ma). In this early stage of development, the existence of oceanic crust, produced by "opening of graben and/or oceans" (Watson 1977) in or between these old cratons, cannot be ruled out. According to Weber & Behr (1983), the rifting process was active until Caledonian time with a maximum in the Ordovician. If so, it must be assumed that it was interrupted rhythmically or episodically by repeated compressive tectogenetic

---

[1]) If an individual Moldanubian (Dalslandian) complex exists within the internides, it is most likely connected with the younger complexes as a succession of incomplete geotectonic cycles.

events, which acted in various parts of the orogene in a different degree. They probably contributed to successive changes of the crustal structure without leading to a consolidation prior to the Caledonian-Variscan time.

References

Ahrendt, H., J. C. Hunziker, and K. Weber (1978): K/Ar-Altersbestimmungen an schwachmetamorphen Gesteinen des Rheinischen Schiefergebirges. Z. dt. geol. Ges. 129, 229–247.

Anderle, H.-J. and S. Meisl (1977): Überblick über den Südtaunus. Exkurs.-führer Geotagung '77 Göttingen, I, 66–76.

Andrusov, D. and O. Corná (1976): Über das Alter des Moldanubikums nach mikrofloristischen Forschungen. Geol. Práce, Spr. 85, 81–89.

Bankwitz, E. and P. (1975): Zur Sedimentation proterozoischer und kambrischer Gesteine im Schwarzburger Antiklinorium. Z. geol. Wiss. 3, 1279–1305.

Bankwitz, P. and E. (1982): Zur Entwicklung der Erzgebirgischen und der Lausitzer Antiklinalzone. Z. angew. Geol. 28, 511–524.

Bankwitz, P. and E., A. Frischbutter, and H.-U. Wetzel (1979): Zu einigen Fragen der Krustenentwicklung in Mitteleuropa. Z. geol. Wiss. 7, 1081–1089.

Behr, H.-J. (1978): Subfluenzprozesse im Grundgebirgsstockwerk Mitteleuropas. Z. dt. geol. Ges. 129, 283–318.

Brause, H. (1969): Das verdeckte Altpaläozoikum der Lausitz und seine regionale Stellung. Abh. dt. Akad. Wiss. Berlin, Kl. Bergb., Hüttenw., Montangeol. 1968, 1, 143 pp.

Brause, H. (1970): Ureuropa und das gefaltete sächsische Paläozoikum. Ber. dt. Ges. geol. Wiss. A, 15, 327–367.

Brause, H. (1979a): Schelfkruste und Drift – konsequente Fortsetzung der v. Bubnoffschen Aussagen. Z. geol. Wiss. 7, 183–191.

Brause, H. (1979b): Mobilistische Aspekte zur Zonengliederung des mitteleuropäischen variszischen Tektogens. Z. geol. Wiss. 7, 1113–1127.

Brause, H. (1980): Differentialmobilismus. Z. geol. Wiss. 8, 405–414.

Burmann, G. (1973): Das Ordovizium der nördlichen Phyllitzone, Teil II: Wippraer Zone. Z. geol. Wiss., them. vol. 1, 9–43.

Chaloupský, J. (1966): Kaledonská a variská orogeneze v ještědském krystaliniku. Sbor. geol. věd, G, 10, 7–37.

Chaloupský, J. (1978): The Precambrian tectogenesis in the Bohemian Massif. Geol. Rdsch. 67, 72–90.

Dornsiepen, U. F. (1979): Rb/Sr Whole Rock Ages within the European Hercynian. A Review. Krystalinikum 14, 33–49.

Douffet, H. (1970): Stratigraphie und Lagerung des Ordoviziums im Bereich der Südvogtländisch-Westerzgebirgischen Querzone. Exkurs.führer „Altpaläozoikum und Vorpaläozoikum des Thüringisch-Vogtländischen Schiefergebirges", Dt. Ges. geol. Wiss. Berlin, 4–5.

Dunning, F. W. and J. Watson (1977): Über die mögliche Erstreckung der Osteuropäischen Tafel bis England und Wales. Z. angew. Geol. 23, 465–470.

Fourmarier, P. and P. Michot (1964): Belgique. In: Tectonique de l'Europe, Moscou, 202–209.

Franke, W., W. Eder, W. Engel, and F. Langenstrassen (1978): Main aspects of Geosynclinal Sedimentation in the Rhenohercynian Zone. Z. dt. geol. Ges. 129, 201–216.

Frischbutter, A. (1982): Zur präkambrischen Entwicklung der Elbezone. Z. angew. Geol. 28, 359–366.

Hirschmann, G. (1966): Assyntische und variszische Baueinheiten im Grundgebirge der Oberlausitz. Freiberger Forsch.h. C 212, 146 pp.

Hirschmann, G. (1975): Regionalbau Lausitz – Dolny Śląsk II, Anteil DDR: Südostteil der Lausitzer Antiklinalzone. Exkurs.führer 22. Jahrestag. Ges. geol. Wiss. DDR, B, 25–32.

Hirschmann, G. (1978): Zu den Möglichkeiten der Korrelation des Proterozoikums zwischen DDR und ČSSR. Berg- und Hüttenmänn. Tag Freiberg 1978, Referate B II, Koll. 7, 9–11.

Hirschmann, G., K. Hoth, and W. Lorenz (1968): Die sedimentologisch-tektonische Entwicklung im Proterozoikum und frühen Paläozoikum der Saxothuringisch-lugischen Zone. XXIII. Int. Geol. Congr. Praha, 4, 141–155.

Hofmann, A. and H. Köhler (1973): Whole Rock Rb-Sr Ages of Anatectic Gneisses from the Schwarzwald, SW Germany. N. Jb. Miner. Abh. 119, 163–187.

Hofmann, J., G. Mathé, J. Pilot, B. Ullrich, and R. Wienholz (1979): Fazies und zeitliche Stellung der Regionalmetamorphose im Erzgebirgskristallin. Z. geol. Wiss. 7, 1091–1106.

Holubec, J. (1966): Stratigraphy of the Upper Proterozoic in the Core of the Bohemian Massif. Rozpr. ČS. Akad. věd, ř. mat. a. přír. věd, 76, 4, 62 pp.

Hoth, K., G. Hirschmann, and W. Lorenz (1970): Das Jungpräkambrium im Bereich der Varisziden und Kaledoniden West- und Nordeuropas und seine Beziehungen zu den paläozoischen Entwicklungsetappen. Ber. dt. Ges. geol. Wiss. A 15, 379—424.

Hoth, K., W. Lorenz, G. Hirschmann, and H.-J. Berger (1979): Lithostratigraphische Gliederungsmöglichkeiten regionalmetamorphen Jungproterozoikums am Beispiel des Erzgebirges. Z. geol. Wiss. 7, 397—404.

Kettner, R. (1946): Some problems of the Algonkian and the Cambrian of Bohemia. Sbor. Stat. geol. Úst. ČSR 13, 41—67 (Engl. summary 59—67).

Kodym, O. (1976): Neue Forschungsergebnisse im Moldanubikum Böhmens. Franz-Kossmat-Symposion. Nova Acta Leopoldina, N. F. 45, 224, 11—22.

Kossmat, F. (1927): Gliederung des varistischen Gebirgsbaues. Abh. Sächs. Geol. Landesamt 1.

Krebs, W. (1977): The Tectonic Evolution of Variscan Meso-Europe. In: Agar, D. V. and M. Brooks: Europe from Crust to Core. Wiley, London, 119—139.

Krebs, W. (1978): Die Kaledoniden im nördlichen Mitteleuropa. Z. dt. geol. Ges. 129, 403—422.

Kvale, A. (1977): Major Features of the European Caledonides and their Development. In: Ager, D. V. and M. Brooks: Europe from Crust to Core. Wiley, London, 81—115.

Lippolt, H. J., H. Schleicher, and I. Raczek (1983): Rb-Sr systematics of Permian volcanites in the Schwarzwald (SW-Germany). Part I: Space of time between plutonism and late orogenic volcanism. Contr. Miner. Petrol. 84, 272—280.

Lorenz, W. and G. Burmann (1972): Alterskriterien für das Präkambrium am Nordrand der Böhmischen Masse. Teil I und II. Geologie 21, 405—433.

Matthews, S. C. (1978): Caledonian Connexions of Variscan Tectonism. Z. dt. geol. Ges. 129, 423—428.

Pacltová, B. (1980): Further micropaleontological data for the Palaeozoic age of the Moldanubian carbonate rocks. Časop. miner. geol., ročn. 25, 275—279.

Paech, H.-J. (1977): Zum Alter tektogener Deformationen im mitteleuropäischen Variszikum. Veröff. Zentralinst. Physik d. Erde Potsdam, 44, 257—280.

Röhlich, P. (1965): Geologische Probleme des mittelböhmischen Algonkiums. Geologie 14, 373—403.

Schmidt, K. (1976): Das „Kaledonische Ereignis" in Mittel- und Südwesteuropa. Franz-Kossmat-Symposion, Nova Acta Leopoldina, N. F. 45, 224, 381—402.

Schwab, M. (1977): Zur geologischen und tektonischen Entwicklung des rhenoherzynischen Variszikums im Harz. In: Probleme der Varisziden in Mitteleuropa und im Gebiet der UdSSR. Veröff. Zentralinst. Physik d. Erde Potsdam. 44,1, 117—147.

Stettner, G. (1974): Probleme des bayerischen Präkambriums. Progr. Int. Corr. Geol., Précambrien des zones mobiles de l'Europe. Conf. Liblice 1972, 109—120.

Stille, H. (1946): Die assyntische Faltung. Z. dt. geol. Ges. 98, 152—165.

Stille, H. (1948): Die kaledonische Faltung Mitteleuropas im Bilde der gesamteuropäischen. Z. dt. geol. Ges. 100, 223—266.

Stille, H. (1951): Das mitteleuropäische variszische Grundgebirge im Bilde des gesamteuropäischen. Beih. Geol. Jahrb. 2.

Svoboda, J. and J. Chaloupský (1966): Crystalline Complexes of the West Sudeten. In: Svoboda, J. et al.: Regional Geology of Czechoslovakia, I, 172—212.

Tollmann, A. (1982): Großräumiger variszischer Deckenbau im Moldanubikum und neue Gedanken zum Variszikum Europas. Geotekt. Forsch. 64, 91 pp.

Urban, L. (1972): Stratigrafické poměry krystalinika v okolí Ličoměřic v Železných horách. Sbor. geol. věd, G 23, 75—112.

Watson, J. (1977): Eo-Europa: The Evolution of a Craton. In: Ager, D. V. and M. Brooks: Europe from Crust to Core. Wiley, London, 81—115.

Weber, K. and H. J. Behr (1983): Geodynamic Interpretation of the Mid-European Variscides. In: Martin, H. and F. W. Eder: Intracontinental Fold Belts. Springer, Berlin-Heidelberg, 427—469.

Wienholz, R., J. Hofmann, and G. Mathé (1979): Über Metamorphose, Tiefenbau und regionale Position des Erzgebirgskristallins. Z. geol. Wiss. 7, 385—395.

Ziegler, P. A. (1982): Geological Atlas of Western and Central Europe. Shell Internationale Petroleum Mij. B. V., vol. 1 and 2.

Ziegler, P. A. (1984): Caledonian and Hercynian Crustal Consolidation of Western and Central Europe — a Working Hypothesis. Geol. en Mijnb. 63, 93—108.

Zoubek, V. (1974): Remarques sur le Précambrien des zones mobiles de l'Europe centrale et occidentale. Progr. Int. Corr. Geol., Précambrien des zones mobiles de l'Europe. Conf. Liblice 1972, 33—72.

Zoubek, V. (1979): Korrelation des präkambrischen Sockels der mittel- und westeuropäischen Variszidzen. Z. geol. Wiss. 7, 1057—1064.

Geochemical Detail Prospecting for Base-Metal and Barite Mineralizations in the Left Rhenish Slate Mountains

M. Krimmel

Geologisches Landesamt Rheinland-Pfalz, Emmeransstraße 36, D-6500 Mainz 1, West Germany

Key Words

Hunsrück
Eifel
Detail-prospection, Metallogenetic, Tectonic and Geological parameters, Contamination, Stream sediment, Soil sampling, Anomalies

Abstract

A reconnaissance prospection for base metal and barite mineralizations in the left Rhenish Slate Mountains by the Bundesanstalt für Geowissenschaften und Rohstoffe, Hannover, in 1973—1977 has identified a large number of promising areas, so that follow-up investigations were carried out within the scope of a research project of the Ministry of Economics and Commerce, Rheinland-Pfalz.

The target in the Hunsrück area has been the range of the Hunsrückschiefer that have accumulated to enormous thicknesses. Besides further base metal deposits known as the "Schieferungsgänge" which are exclusively orientated in SW-NE-direction, synsedimentary, stratiform deposits in local depressions — third order basins — have been searched for.

The geological situation, the geochemistry of the country rocks and the identical Pb-, Zn- and Cu-distribution in the ore and the Hunsrückschiefer suggest a deviation from these shales.

The distribution of the mineralizations in the Eifel area is controlled by NNE-SSW and NW-SE-striking strain structures. Therefore areas of plunging anticlinal axes and horizontal flexure zones, where these structures are mainly installed, have been targeted. Finally 14 anomalies identified by the reconnaissance operations were subjected to geochemical follow-up investigations. The various contaminations in this central-European region strongly affect stream sediment sampling. For that reason the deepest soil-horizons were sampled during the systematic soil-survey and analyzed for Pb, Zn, Cu, Hg and Ba.

Obtained soil-anomalies were supplemented by systematic collection of hard-rock samples. This detailed work confirmed the existance of three Ba anomalies in the west Eifel and two base metal anomalies in the Hunsrück. The Ba anomalies are located in areas of plunging axes at the marginal region of the Eifel N-S-Zone. The base metal anomalies lie within the range of the Hunsrückschiefer, where dispersion halos are developed, which contrast clearly with the normal element distributions.

Kurzfassung

Im Rahmen eines Forschungsprojektes des Min. f. Wirtschaft und Verkehr, Rhld.-Pf. wurde eine Detailprospektion auf Buntmetall- und Barytmineralisationen im linksrheinischen Schiefergebirge durchgeführt, nachdem eine Übersichtsprospektion der BGR, Hannover eine Reihe vielversprechender Anomalien erbracht hatte.
Zielgebiet im Hunsrück war das Verbreitungsgebiet der Hunsrückschiefer, die zu enormen Mächtigkeiten angehäuft worden waren. Neben weiterer Buntmetallvorkommen in der Art der bekannten „Schieferungsgänge", die ausschließlich variscisch streichen, galt die Suche auch synsedimentären, stratiformen Vorkommen, in lokalen Depressionen, sog. Third-order basins. Die geologische Situation, die Geochemie der Sedimente und die übereinstimmende Pb-, Zn- und Cu-Verteilung in den Erzproben und den Hunsrückschiefern läßt vermuten, daß die Buntmetalle aus den Schiefern selbst stammen. Das Auftreten der Mineralisationen in der Eifel ist gebunden an NNE-SSW und NW-SE-streichende Dehnungsstrukturen in den sandigen Sedimenten. Für die Anlage solcher Dehnungsstrukturen sind einmal die Achsenrampen der unter die Eifeler N-S-Zone östl. wie westl. abtauchenden Sättel sowie horizontale Achsenflexurzonen prädestiniert und somit die Zielgebiete in der Eifel. So wurden 14 durch die Übersichtsprospektion indizierten Anomalien weiterer follow-up Untersuchungen unterzogen. Da mit mehr oder weniger starken Kontaminationen, vor allem der Bachsedimente, gerechnet werden mußte, wurden, im Rahmen einer systematischen Bodenbeprobung die untersten Horizonte beprobt und auf die Elemente Pb, Zn, Cu, Hg und Ba analysiert. Für die Überprüfung erhaltener Bodenanomalien und für weitere chemische und petrographische Untersuchungen wurden systematisch hard-rock Proben entnommen. So konnten drei Ba-Anomalien in der Eifel und zwei Buntmetallanomalien im Hunsrück verifiziert werden. Die Ba-Anomalien liegen in Bereichen der Achsenrampen im Randgebiet zur Eifeler N-S-Zone; die Buntmetallanomalien liegen im Verbreitungsgebiet der Hunsrückschiefer. Dort sind Dispersionshöfe ausgebildet, die klar vom normalen Verteilungsmuster der Elemente verschieden sind.

Introduction

After quantitative evaluation of all known ore and mineral deposits, The Rhenohercynian Zone of the European Variscides has to be regarded as a very promising area. According to Routhier (1980) nearly 90% of the Pb-Zn production and reserves in the Federal Republic of Germany descend from parts of the Rhenish Slate Mountains and the Harz. In the same way such a regional concentration in the Rhenohercynicum can be established for the appearance of barite mineralizations. The left-Rhenish part of the Rhenish Slate Mountains, where a great number of Pb-Zn ore and barite mineralizations are known, also represents a promising area. But all these deposits which have been subjected to intensive mining are very close to the surface and have been recognized and developed only by mine digging or outcrops.
In order to attempt to detect possible, yet unknown, concealed accumulations, a reconnaissance prospecting for base metal and barite mineralizations was developed by the Bundesanstalt für Geowissenschaften und Rohstoffe (BGR), Hannover, between 1973 and 1977 in that part of the left-Rhenish Slate Mountains which belongs to Rheinland-Pfalz.

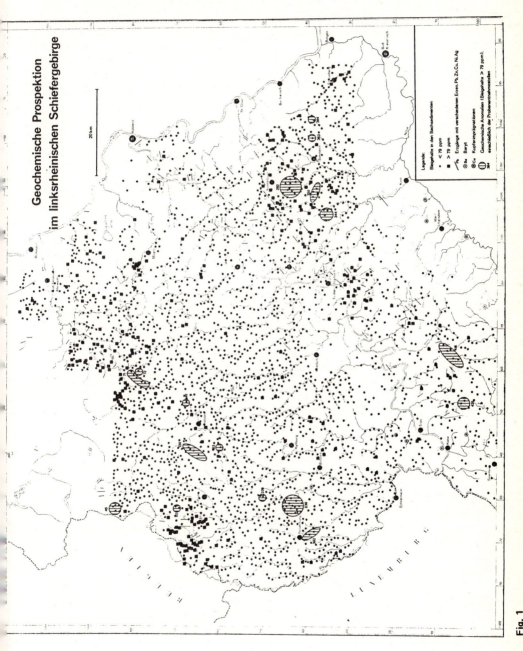

Fig. 1
Geochemical reconnaissance stream sediment prospection in the left Rhenish slate mountains (after Emmermann et al., 1981).

Within an area of $8\,000\,km^2$, $6\,000$ stream sediment samples have been taken and analyzed for Pb, Zn, Cu and Ba. This prospecting led to the discovery of 200 anomalies regarded as uncontaminated and significant (Fig. 1). Of these, 28 anomalies were chosen by a mathematical treatment by the BGR, Hannover and recommended for further follow-up investigations (Emmermann et al., 1981) within the scope of a research project by the Ministry of Economics and Commerce, Rheinland-Pfalz in cooperation with the Bundesanstalt für Geowissenschaften und Rohstoffe. Of these recommended anomalies, the most promising targets have to be defined for beginning effective detailed follow-up works.

General Ore Control Parameters

Hunsrück

The detailed search for concealed mineralizations must rely more and more on a practical model, which should emphasize the geological, tectonic and geochemical parameters considered favorable for the genesis, distribution or character of a mineralization within a particular geological environment. Based on the geochemical surface indications, the prospecting model should yield special leads to possible, concealed accumulations of base metal or barite within this Paleozoic deposition environment.

In the Hunsrück area the character and distribution of the known Pb-Zn deposits show the following metallogenetic regularities:
- They are exclusively located within the range of the Hunsrückschiefer, south of the Boppard-Dausenauer overthrust (Fig. 2).
- They occur in SW-NE striking, particular vein systems (Fig. 2). The mineralizations follow the cleavage, mostly within the sandy parts of the shales and form lense-shaped, SE-dipping echeloned ore zones (Herbst & Müller, 1966; Cup, 1955).
- The mineralizations are of oligometallic character. That means (see Table I) that the amounts of Pb, Zn and Cu are rather low in the Hunsrück ores. A comparison of these values with the Pb-, Zn- and Cu contents of the Hunsrückschiefer in a variation diagram (Fig. 3) (Krimmel, 1984) shows a striking overlap of these both fields of distribution. Recent geochemical and isotopic data suggest that the metals could have been derived from the shales, as pointed out by many authors for the origin of most mineralizations in the Rhenohercynicum (Nielsen, 1966; Walther, 1982; Möller et al., 1979; amongst many). Based on that, newer physicochemical data and results in combination with the geological and tectonic features of that part of the Rhenohercynicum might establish a similar model for the origin and setting of the mineralizations in the Hunsrück (Krimmel, 1984).

Interpretation of paleogeographic, lithologic, tectonic and geochemical data also yielded ore control parameters, which could have been responsible for the formation of synsedimentary, stratiform enrichments.
- During the Devonian the Hunsrück area has been part of the troughfacies of the Variscan geosyncline within the Rhenohercynian zone (Fig. 4). The trough itself has been subdivided into "Schwellen" and "Becken" with different deposition environments.

These "Becken" (second-order basins) were filled with pelagic dark shales, deposited during continuous synsedimentary subsidence and locally built up to enormous thicknesses (up to $6\,000\,m$; Mittmeyer, 1980).
- A possible formation of regional depressions (third-order basins) which could have played a role as geomorphic traps for the sulfides is indicated by black-shale layers locally with 5 to 10 times higher concentrations in typical elements like V, Co, Zn,

Pb, Cu, Ni and Fe (Fig. 5). The occurrence of diagenetic pyrite and of pyritic fossils is typical for such shales.

— Synsedimentary tectonic activities, that could have triggered the formation of third-order basins, are suggested by the displacement of the Soonwald-Schwelle (Meyer & Stets, 1980) and from examination of satellite photos (Kronberg, 1979).

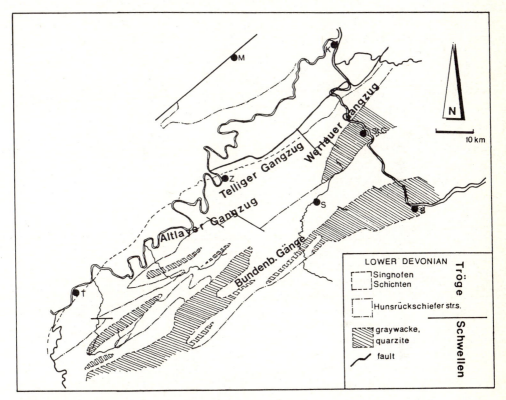

Fig. 2
Geological and paleogeographical sketch and the positions of ore veins in the Hunsrück (after Mitt-meyer, 1980; Weiler, Geib, Mittmeyer, 1966).

Table I: Ore samples from Theodor mine (T1—T3), Eid mine (Eid) and Barbarasegen mine (Altlay) of the Hunsrück area.

Sample Nr.	Zn (%)	Pb (%)	Cu (%)	Fe (%)	Co (ppm)	Ag (ppm)	Mn (ppm)	Cr (ppm)	V (ppm)	Mo (ppm)	Ni (ppm)
T 1	25	14	0,4	3.7	73	230	300	2	13	10	17
T 2	26	14	0.5	3.7	32	220	292	3	11	4	19
T 3	27	18	0,5	3.7	41	241	425	23	11	5	28
Eid	12	18	0,4	0.8	81	144	72	12	—	33	20
Altlay	0.9	0,7	29	29	13	180	345	22	—	39	20

— Bimodal volcanism and anomalous high and differentiated heat-flow during the synclinal stage of the Rhenohercynicum could be indicative of an epicontinental marginal basin. Most of the synsedimentary deposits, as Tom, Sullivan or McArthur River, are restricted to such epi- or intracontinental rifted basins (van den Boom et al., 1980).

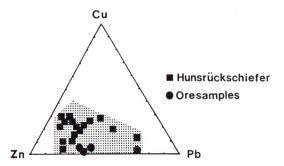

Fig. 3
Pb-, Zn-, Cu-variation diagram of the Hunsrückschiefer- and ore-samples.

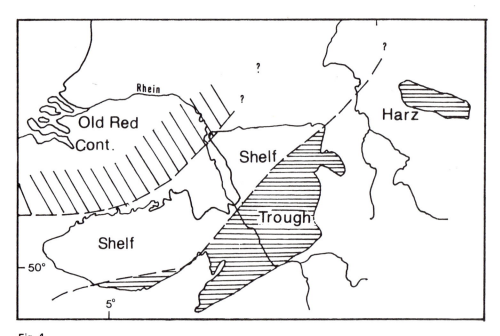

Fig. 4
Paleogeographic situation in the middle European region during the Devonian (after van den Boom, 1980).

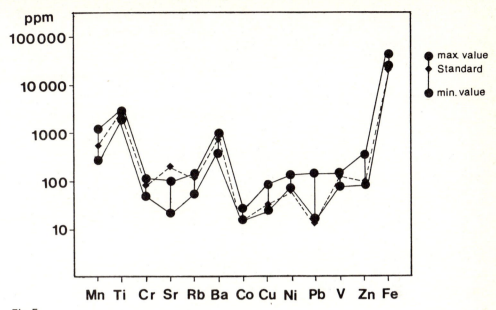

Fig. 5
Minimum and maximum contents of trace elements in the Hunsrückschiefer in comparison to a standard mean (standards: Rösler, Lange, 1976; Rose, Hawkes, Webb, 1980).

Because of these parameters, anomalies located in the range of the Hunsrückschiefer
have been regarded as targets for further follow-up works. Fig. 6 and Table II show the
investigated anomalies in the Hunsrück area. The given element concentrations in
Table II originate from the reconnaissance prospection by the BGR, Hannover.

Eifel

The Eifel represents the former shelf region of the Variscan geosyncline, with sandy, nearshore, shallow sea deposits (Fig. 4). In this area the ore controls have the following characteristics:
— The occurrence of known mineralizations is restricted to NNE-SSW and NW-SE
  striking strain structures whithin these sandy sediments (Richter, 1963).
— Of great importance are horizontal flexure zones resulting from a slicing of lower
  tectonic levels during compression, and which created deep-reaching fissures and
  feather joints. These structures could have channeled mineral solutions, such as the
  well-known barite mineralization at Uersfeld/Eifel (Weisser, 1963; Fig. 7).
— The marginal region east and west of the Eifel N-S-Zone represents a region of special
  tectonic setting with a system of plunging axes underneath the N-S-Zone. This vertical
  stretching of the rock series occasionally is overlapped by horizontal flexuring which
  enhances the fracturing and promotes the ascent of mineral solutions. These targets
  for further geochemical investigations in the Eifel are shown in Fig. 8 and Table II.

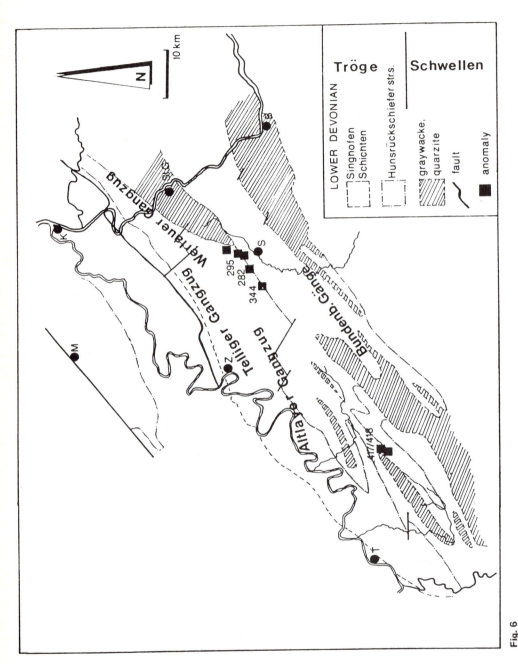

Fig. 6

Location of the significant and investigated anomalies in the Hunsrück area.

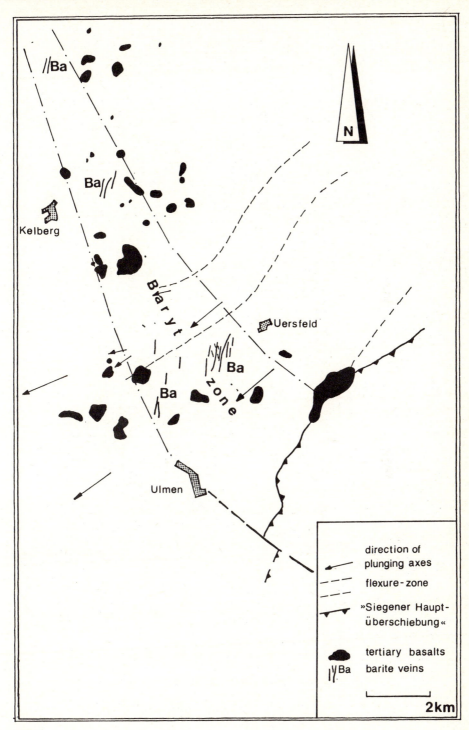

Fig. 7
The barite veins at Uersfeld/Eifel. A tectonic sketch after Weisser, 1963.

Table II: The investigated anomalies in the Eifel and Hunsrück.

Anomaly Nr.	Map Nr.	Strat. Unit	Sampling Procedure	Anomalous Elements (ppm)
71	5606	Klerfer Sch., tu	SS, HRS	Pb-(Ba) Pb: 85–170 Ba: 275
72	5606/5607	Klerfer Sch., tu	SS	Ba-Zn–(Pb) Ba: 200–310 Zn: 280–325 Pb: 80
182	5805	Klerfer Sch., tu	SS	Pb Pb: 80
183	5805	Klerfer Sch., tu	SS	Ba-(Cu-Zn-Pb) Ba: 180–320 Cu: 40, Zn: 300 Pb: 150
185	5805	Klerfer Sch., tu	SS	Ba Ba: 270
192	5805	Klerfer Sch., tu	SS	Ba Ba: 200–260
195	5806	Klerfer Sch., tu	SS	Ba-Zn Ba: 180–360 Zn: 370–460
226	5903/6003	Klerfer Sch., tu	SS	Ba-(Pb-Zn-Cu) Ba: 200–1150 Cu: 40–55 Pb: 300, Zn: 950
229*	5904	Klerfer Sch., tu	SS, HRS	Pa(Pb-Zn-Cu) Ba: 185–980 Cu: 40–55 Pb: 130, Zn: 660
230*	5904	Klerfer Sch., tu	SS, HRS	Ba-(Pb-Zn) Ba: 180–1650 Pb: 100, Zn: 285
282*	5910	Hunsrückschiefer i.w.S., tu	SS	Pb-Cu-Zn-Ba Pb: 80–320 Zn: 290–400 Cu: 50, Ba: 300
295	5911	Hunsrückschiefer i.w.S., tu	SS	Pb-Zn Pb: 85–880 Zn: 300–320
344*	6009/6010	Hunsrückschiefer i.w.S., tu	SS	Zn-Pb-(Ba) Zn: 290–2700 Pb: 80–170
417/418	6027	Hunsrückschiefer i.w.S., tu	SS, HRS, WS, SSS	Pb Pb: 95–150

*   = in cooperation with the BGR, Hannover.
SS   = soil sampling, SSS = stream sediment sampling, WS = water sampling
HRS = hard-rock sampling

Contamination Problems

A main problem in carrying out geochemical exploration in a populated area, such as West-Germany, is the contamination of stream sediments by urban and industrial waste. In addition, in the Eifel and Hunsrück areas the waste heaps and culverts of the former intensive Pb-Zn minings are also likely to affect water and stream sediment pollution. Evaluation of the intensity and influence of such man-made pollutions, in order to recognize contaminated and uncontaminated environments, is of great importance for the assessment of geochemical data, but in most cases it is rather complex.

Tests being carried out during the prospection showed that selective extraction procedures and investigations of a distinct spectrum of heavy metals could yield practical hints for comparing contaminated and non-contaminated areas. It appears that relatively strong overprints of the sediments, mainly Zn, Pb and Hg in weakly adsorbed forms, are narrowly localized in areas of high population density and somewhat broader (in typical and significant enrichments of Pb, Zn and Cu and high correlation of Pb, Zn, Cu, Hg and Fe) in areas of former ore mining. These results showed that mainly the elements Zn, Pb, Hg and Cd, weakly adsorbed and highly enriched, are the most diagnostic elements for urban and industrial pollution (Krimmel, 1984).

Because of such problems, sampling of stream sediments was omitted. Moreover, stream sediments got lost in areas where the creek beds were cased or embedded for a compact sampling. For an efficient evaluation of geochemical data, soil sampling grids have been installed with respect to the varying morphologic and geologic features within the investigation areas. The samples were taken from the deepest horizons, mostly B and C horizons, and analyzed for Pb, Zn, Cu in the hot extractable form (semi-conc. HNO_3) and Ba, Hg as total amounts. For reviewing anomalous soil-sample profile sections and for detailed chemical and petrographic investigations (the last is not evaluated in this summary), hard rock samples have also been collected.

Significant Anomalies

Eifel

In the Eifel area the Ba anomalies 226, 229 and 230 were confirmed and their localization improved. These significant anomalies are located in the range of a SW-NE striking area at the marginal region of the Trier-Bitburger-Mulde (Fig. 8).

In the anomaly 226 the highest Ba contents of the soils, up to 1000 ppm (background: 500 ppm), are concentrated within an area of 3 km². Further detailed soil sampling on parallel profiles is needed to refine the source of the indicated anomaly.

In anomaly 229 the analyses of systematically collected rock samples could improve the Ba values in the soil, which have yielded a N-S striking halo with amounts up to 1400 ppm. These amounts lie within an area of a horizontal flexure zone.

The most significant anomaly in the Eifel is represented by the Ba anomaly 230 (Fig. 9). High Ba contents occur SW of Merkeshausen, N and W of the Burbesberg and NE of Altscheid where amounts up to 5000 ppm Ba (background: 270 ppm) could be found in the soils (Fig. 9). The analyses of rock samples, which were collected parallel to the soil profiles, yielded the same distribution pattern of anomalous Ba amounts as in the soils. Amounts up to 4800 ppm Ba could be found in the rocks.

By comparing the distribution of the anomalous ranges (Fig. 9), where the highest Ba contents are concentrated, dispersion halos become recognizable along different profile sections. These halos might be related to strain fissures in the range of the plunging

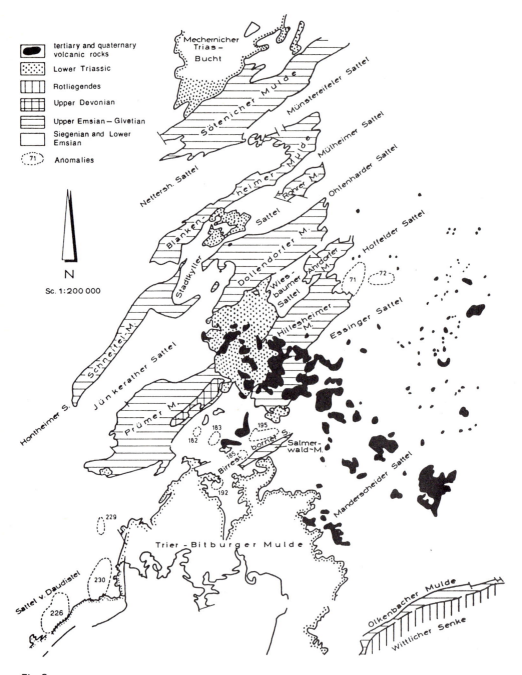

Fig. 8
Location of the significant and investigated anomalies in the Eifel area.

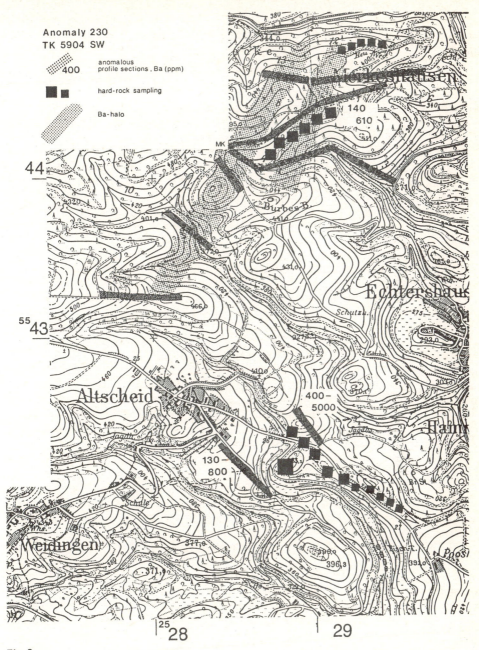

Fig. 9
Position of the anomalous profile sections in anomaly 230 and the points of hard-rock sampling.

Daudistel-Scheuerndell anticline overprinted by horizontal flexuring (Leppla, 1905). In this anomaly further exploratory excavations and possibly prospecting shafts should follow.

Hunsrück

In the Hunsrück area the Pb-Zn anomalies 344 and 282, indicated by anomalous stream sediments, could not be confirmed by soil sampling. In contrast the extreme Pb and Zn amounts appeared to be anthropogenous contamination.
The Pb anomaly 295 at Maisborn, however, must be regarded as a significant anomaly. The evaluation of the soil-sample grid shows a Pb-Zn halo striking NE-SW and is marked by Pb amounts up to 600 ppm and Zn contents from 130 to 1 100 ppm. The local background of these elements lies at 70 and 90 ppm (Fig. 10). SW of anomaly 295, a SW-NE extending region is marked by anomalous Pb concentrations in stream sediments (Fig. 11). Further geochemical investigations are needed to determine whether this area in fact represents a continuation of anomaly 295 in a southwestern direction.
The detailed geochemical investigations in the anomaly 417/418 in the West-Hunsrück yielded two anomalous areas. First the region of the headwater of the Krennerichbach, where heavy metal-bearing waters emerge within a water pollution control zone and therefore are not affected by any contamination (Fig. 12). Water samples show Pb values of 30 ppb, Mn contents up to 300 ppb, and Fe concentrations in the range 600−23 000 ppb. In the stream sediments the Pb contents reach values of 3 000 ppm, with a local background of 150 ppm; Fe is enriched in some samples up to 15 %, whereas the background is 2 %. West of Beuren a Pb dispersion halo tending NNE-SSW could be found by a detailed soil sample grid (Fig. 13). The Pb concentrations in the soils over Hunsrückschiefer reach values up to 700 ppm, with a local background of 90 ppm, whereas Zn and Cu show no significant enrichments along the soil profiles. Slightly higher Zn amounts at the deeper slope are triggered by secondary enrichments in the colluvial hill waste.
These anomalies in the western part of the Hunsrück area should be reviewed by geophysical measurements to confirm the geochemical indications.

Conclusions and Prospects

Reconnaissance and detailed prospection in the left Rhenish Slate Mountains yielded significant geochemical indications within favorable geological environments. Interpretation of the paleogeographic, lithologic, tectonic and geochemical data yielded ore control parameters which further determine the targets in that part of the Rhenohercynicum.
Selective analyzing procedures showed that numerous sources of contamination strongly affect geochemical exploration of stream water and -sediment in such populated areas. Therefore selective sampling of the deepest soil horizons and hard rocks has been preferred. A task of further prospection works should be to check the geochemical indications by geophysical measurements or even prospecting shafts.

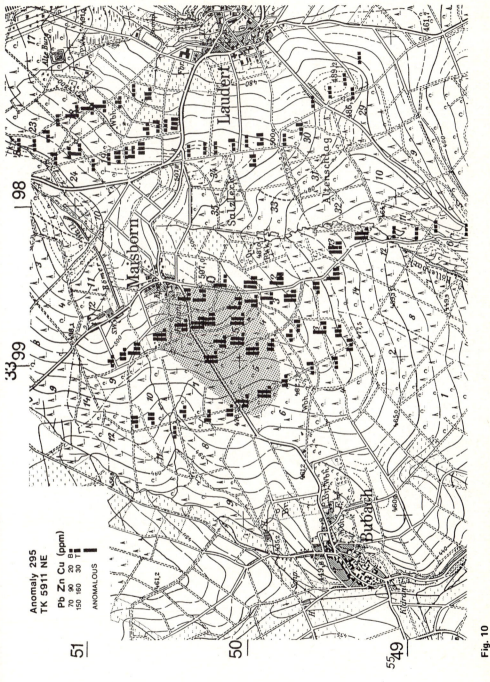

Fig. 10
Map of anomaly 295: the distribution of Pb, Zn and Cu.

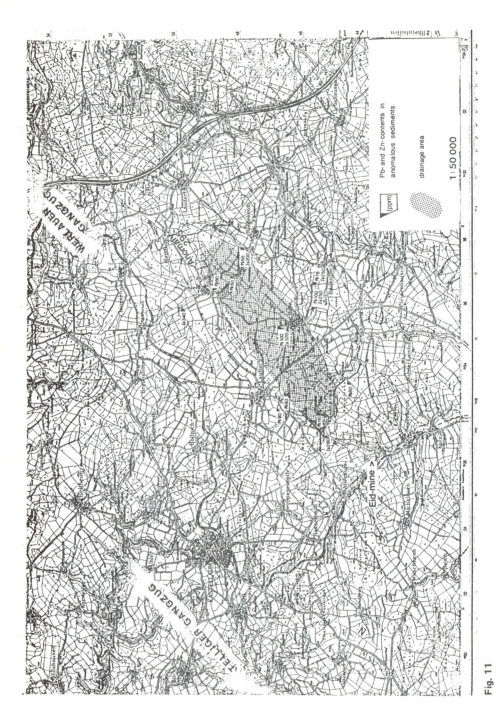

Fig. 11

Stream sediment anomaly SW of Pb anomaly 295.

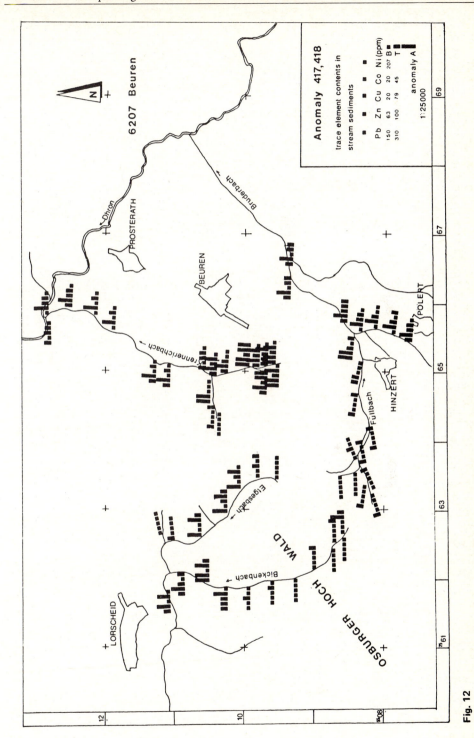

Fig. 12
Distribution of Pb, Zn, Cu, Co and Ni in stream sediments in anomaly 417/418.

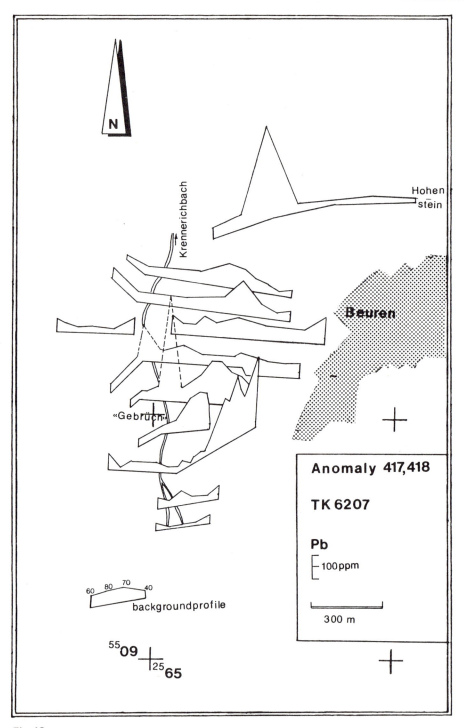

Fig. 13
Pb contents of the soils W of Beuren/Hochwald.

References

Cup, C. (1955): Tectonic and genesis of the lead-zinc-ores of Tellig, Hunsrück, W.-Germany. Geologie en Mijnbouw, N. Serie 17, 285—318.

Emmermann, K.-H., H. Fauth, R. Hindel, Chr. Ree (1981): Geochemische Übersichtsprospektion im linksrheinischen Schiefergebirge. Erzmetall, 3, 152—159.

Herbst, F., H.-G. Müller (1966): Der Blei-Zink Erzbergbau im Hunsrückgebiet. Gewerkschaft Merkur 48 S.

Krimmel, M. (1984): Geochemische Detailprospektion auf Buntmetall- und Barytmineralisationen im linksrheinischen Schiefergebirge. 200 S., 49 Abb., 19 Tab., Diss. [unveröff.], Mainz, 1984.

Kronberg, P. (1976): Bruchstrukturen des Rheinischen Schiefergebirges, des Münsterlandes und des Niederrheins. Kartiert in Aufnahmen des Erkundungssatelliten ERTS-1. Geol. Jb., A 33, 37—48.

Leppla, A. (1905): Erläuterungen zu Blatt Waxweiler.

Meyer, W., J. Stets (1980): Zur Paläogeographie von Unter- und Mitteldevon im westlichen und zentralen Rheinischen Schiefergebirge. Ztschr. dtsch. geol. Ges., 126, 15—29.

Mittmeyer, H.-G. (1980): Zur Geologie des Hunsrückschiefers. Natur und Museum, 110, 148—155.

Möller, P., G. Morteani, G., J. Hoefs, P. P. Parekh (1979): The origin of the ore-bearing solutions in the Pb-Zn veins of the western Harz, Germany, as deduced from rare-earth element and isotope distributions in calcites. Chem. Geol., 26, 197—215.

Nielsen, H., (1968): Schwefel-Isotopen aus St. Andreasberg und anderen Erzvorkommen des Harzes. N. Jb. Min. Abh., 109, 289—321.

Richter, M. (1963): Die Blei-Zink-Erzvorkommen in der Umgebung von Virneburg, Südeifel. Erzmetall, 16, Heft 12.

Rösler, H.-J., H. Lange (1976): Geochemische Tabellen. Ferd. Enke Verlag, 674 S., 2. Aufl., Stuttgart.

Rose, A. W., H. E. Hakes, J. S. Webb (1979): Geochemistry in mineral exploration. Academic Press, 654 S., London, Toronto, New York.

Routhier, P. (1980): Ou sont les métaux pour l'avenir? Les provences métalliques. Essai de métallogenie globale. Mém. BGRM 105, 410S., Orleans.

Schenk, E. (1937): Die Tektonik der Mitteldevonischen Kalkmuldenzone in der Eifel. Geol. Jb., 58, 2—35.

Van den Boom, G. et al. (1980): Stratiform Cu-Pb-Zn deposits. Geol. Jb., D, Heft 40, 201 S.

Walther, H. W. (1982): Die variszische Lagerstättenbildung im westlichen Mitteleuropa. Ztschr. dtsch. geol. Ges., 133, 667—686.

Weiler, H. (ed.) (1976): Erläuterungsbericht des wasserwirtschaftlichen Rahmenplanes für das Moselgebiet in Rheinland-Pfalz.

Weisser, D. (1963): Tektonik und Barytgänge in der SE-Eifel. Ztschr. dtsch. geol. Ges., 115, 33—68.

Karten und Berichte des Niedersächsischen Landesamtes für Bodenforschung und der Bundesanstalt für Geowissenschaften und Rohstoffe Hannover, Nr. 72: Geochemische Übersichtsprospektion im linksrheinischen Schiefergebirge, 1979 und Nr. 85: Geochemische Detailprospektion im linksrheinischen Schiefergebirge, 1980.

Maps

Geologische Übersichtskarte Rheinland-Pfalz 1 : 250 000.

Übersichtskarte der Bodentypen-Gesellschaften 1 : 250 000.

Geologische Übersichtskarte der Eifel 1 : 200 000.

Übersichtskarte von Rheinland-Pfalz 1 : 250 000 und 1 : 500 000.

Geol. Karte 5508 Kempenich

Geol. Karte 5904 Waxweiler

Geol. Karte 6003 Mettendorf

Geol. Karte 5903 Neuerburg

Geol. Karte 6207 Beuren/Hochwald

TK 5508 Kempenich	TK 5903 Neuerburg
TK 5609 Mayen	TK 6003 Mettendorf
TK 5608 Virneburg	TK 5911 Kisselbach
TK 5607 Adenau	TK 5910 Kastellaun
TK 5606 Üxheim	TK 6010 Kirchberg
TK 5806 Daun	TK 6009 Sohren
TK 5805 Mürlenbach	TK 6207 Beuren/Hochwald
TK 5904 Waxweiler	TK 6506 Trier-Pfalzel

Aims and Status of Establishing a Gravity Fixpoint Field in the Federal Republic of Germany

E. Czuczor / P. Lux / R. Strauß

Hessisches Landesvermessungsamt, Schaperstr. 16, D-6200 Wiesbaden

Key Words

First, second and third order gravity networks

Abstract

Surveying authorities of the Federal Republic of Germany are presently establishing a gravity fixpoint field. Measuring first-order gravity net (Hauptschwerenetz der Bundesrepublik Deutschland — DHSN 82) with 299 stations was finished in 1982. The evaluation of the measurements has proceeded as far as statements about accuracy are possible. Already measurements of second-order gravity nets have been finished by some federal states. For the third-order gravity nets, a density of 1 fixpoint per $5\,\mathrm{km}^2$ is provided. The time required until third-order completion will also depend to what extent other geosciences want to use these data.

Aims

The surveying authorities of the states of the Federal Republic of Germany are creating a gravity fixpoint field, which has the following aims:
a) Preparation of the gravity fixpoints as reference points for following gravity measurements of other users of gravity methods.
b) Preparation of the results of gravity measurements for geological interpretation.
c) Representation of the gravity field by gravity anomalies for geological, geophysical and geodetical tasks.
d) Evaluation of geoid undulations and plumb-line deflections for reduction of geodetic measurements to the ellipsoid.
e) Evaluation of potential differences and orthometric heights from the combination of levelling and gravity measurements.
f) Evaluation of secular and long-periodic gravity variations by repeated measurements; thereby eventually parallel proving height changes which result from levelling.
The points of the gravity net shall be marked durably and shall be measured with an accuracy of $10 \times 10^{-8}\,\mathrm{ms}^{-2}$.
This article is to give an impulse for a closer cooperation of geodesists and other geoscientists in the field of gravimetry.

Actual Hierarchy of Nets

Gravity Base Net 1976 of the Federal Republic of Germany
(Schweregrundnetz 1976 der Bundesrepublik Deutschland — DSGN 76)

The German Geodetic Commission (DGK) in 1974 to 1976 created this net as a new gravity base net.
It is characterized by:
a) only 21 gravity stations equally distributed in the Federal Republic, which makes possible rapid measurement and easier supervision; every station consists of one center and three excenters;
b) a number of stations identical with points of the former gravity net and the I.G.S.N.71;
c) all stations being situated in buildings to minimize the influences of temperature, microseismic and mass variations;
d) all gravity differences being measured independently twice with the same four La Coste and Romberg model-G gravity meters;
e) niveau and scale taken from four absolute gravity measurements carried out in 1977 at the stations München, Wiesbaden, Braunschweig and Hamburg.
The r.m.s.e. (random mean square errors) of the adjusted gravity values lie between 6 and $11 \times 10^{-8} \, \mathrm{ms}^{-2}$.

First Order Gravity Net 1982 of the Federal Republic of Germany
(Hauptschwerenetz 1982 der Bundesrepublik Deutschland — DHSN 82) (Fig. 1)

The surveying authorities in 1978 began establishing the DHSN 82. Measurements were finished in 1982; since then adjustment is performed by the Hessian land surveying office (Hessisches Landesvermessungsamt) as a computing center.
With a density of about 1 fixpoint per $1000 \, \mathrm{km}^2$, the net consists of 299 gravity fixpoints; 21 of them are identical with the gravity base net stations. The points have been chosen only from the durable marked fixpoints of the official trigonometric and levelling networks. The elated points are especially suitable for gravity measurements and lie in presumably stable areas.
The gravity differences were measured in the manner A-B B-A with at least two gravity meters; altogether 15 model-G and model-D LaCoste-Romberg gravity meters were used.
At the present state, adjustment gives r.m.s.e. of 3 to $5 \times 10^{-8} \, \mathrm{ms}^{-2}$ for the adjusted gravity values.

Gravity Nets of Second and Third Order

To make available gravity data of dense distribution in the above-mentioned modern and precise system, surveying authorities are creating second- and third-order gravity nets.
In the second-order gravity net, density of points amounts to 1 fixpoint per $100 \, \mathrm{km}^2$; in the third-order gravity net it amounts to 1 fixpoint per $5 \, \mathrm{km}^2$.
The rules of selection of points are the same as in first-order net.
Establishing the second-order net is not yet finished in all federal states; the Hessian second-order gravity net may give an example for a completed net (Fig. 2). The attained r.m.s.e. of the adjusted gravity values in this net are 3 to $4 \times 10^{-8} \, \mathrm{ms}^{-2}$.

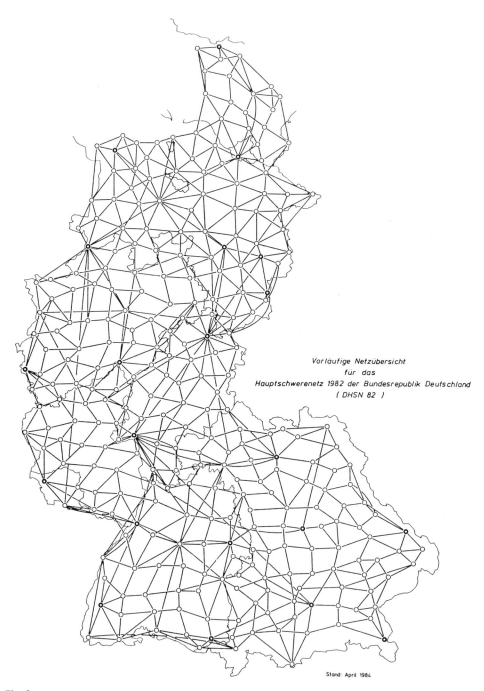

Vorläufige Netzübersicht
für das
Hauptschwerenetz 1982 der Bundesrepublik Deutschland
(DHSN 82)

Stand: April 1984

Fig. 1
Preliminary net overview of the 1982 first-order gravity net in the Federal Republic of Germany

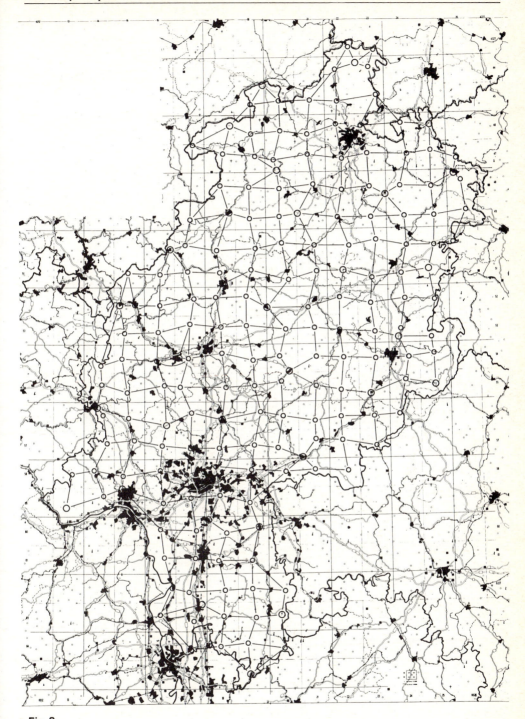

Fig. 2
Hessian second-order gravity net

To build up third-order gravity nets for the present, gravity measurements along the lines of the main levelling net are carried out. A completed net, which will cover a whole area, may have the same structure as the second-order net shown above. In the federal state of Hessen it is planned to complete third-order net in whole areas following geologic interests; there will be a cooperation with the Hessian geological office.

Former and Other Gravity Nets in the Territory of Federal Republic

The modern and precise gravity nets presented above are not the first activities in this field.

In chronologic development local nets for prospecting purposes, which were established by private companies (Prakla, Seismos), are to be mentioned first.

A first measurement of gravity at the earth's surface all around the (former) German territory began in 1934 (Geophysikalische Reichsaufnahme) and still continued after the end of the war. Niveau and scale were taken from a net of pendulum points (a first base net). Accuracy of the gravity values is better than $500 \times 10^{-8}\,\mathrm{ms}^{-2}$.

A renewed gravity base net was established from 1952 till 1958 by the German Geodetic Commission (DGK), a gravity meter net which also took niveau and scale from pendulum points. It was provided to participate in a first-order world gravity net, which has been started at the same time.

In 1971 this world gravity net with name I.G.S.N. 71 was installed by the I.U.G.G., as an obligatory world-wide reference system. In the area of the Federal Republic, r.m.s.e. lie between 15 and $45 \times 10^{-8}\,\mathrm{ms}^{-2}$.

Comparison Between Old and New Gravity Nets

I.G.S.N. 71 is not replaced by the new gravity base net DSGN 76; yet I.U.G.G. has acknowledged DSGN 76 as a new equivalent reference system. Through DSGN 76 the new first- to third-order gravity nets are referred to modern high-precision absolute gravity measurements, which is remarkable progress besides the higher accuracy.

The new modern gravity nets are superior in two ways to all older gravity nets mentioned above:

a) because of their precision, which mainly refers to instrumental progress;

b) because of the use of durably marked fixpoints, by which the gravity nets take advantage of the well known preservation of the official geodetic nets.

On Utilization of the New Gravity Nets

The gravity values of the new gravity fixpoint field are available to everyone from land surveying offices.

Contacts between surveying and geological authorities about performing gravity measurements and elaborating charts of anomalies already took place in 1975. Meanwhile the surveying offices were engaged in establishing the necessary nets of higher order. For the activities in the gravity fixpoint field to proceed, it would be useful if other geosciences would name their requirements concretely.

The measurements in third-order nets will take many years. In most federal states, those measurements have not yet begun. The high costs will only be met when a suitable utilization of the data is in view.

References

Apel, H. (1976): Aufbau eines Schwerefestpunktfeldes als Aufgabe der Landesvermessung. Zeitschrift für Vermessungswesen Nr. 8, S. 359.

Bartsch, E. (1979): Der Aufbau des Schwerefestpunktfeldes in der Landesvermessung. Mitteilungen des Landesvereins Hessen des Deutschen Vereins für Vermessungswesen Nr. 1.

Cannizzo, L., Cerutti, G., Marson, I. (1978): Absolute Gravity Measurements in Europe. Il Nuovo Cimento Vol. 1 c, N. 1.

Kneißl, M. (1962): Literatur über das Europäische Gravimeter-Eichsystem. Deutsche Geodätische Kommission, München.

Morelli, C. et al. (1971): The International Gravity Standardization Net 1971 (I.G.S.N. 71). IUGG/IAG Publication Speciale No 4, Bureau Central de l'Association Internationale de Geodesie, Paris.

Sigl, R. et al. (1981): Das Schweregrundnetz 1976 der Bundesrepublik Deutschland (DSGN 76). Deutsche Geodätische Kommission, Reihe B, Nr. 254, München.

Variscan Thrust Tectonics in the Bohemian Massif — First Results Revealed by Reflection Seismology

Č. Tomek / I. Ibrmajer / K. Cidlinský

Geofyzika Brno, P. O. Box 62, 61246 Brno, Czechoslovakia

Key Words

Variscan orogen
Bohemian Massif
Reflection seismology
Thrust tectonics
Crustal Structure

Abstract

Seismic reflection data collected from the eastern part of the Bohemian Massif (B. M) provide information on Late Paleozoic Variscan compressional tectonics. Those data show a series of remarkably continuous northwest-dipping reflectors that extent about 20 km beneath the eastern margin of the Bohemian Massif and can be traced as deep as 10 km. Seismic reflection lines across the Rhenohercynian folded Culm Flysch zone and Moldanubian zone of the East Bohemian Massif (easternmost European Variscides) suggest that the upper ten kilometers of the crust in this region is composed of tectonically thickened metamorphic rocks of the Moldanubicum, Paleozoic sediments and basement thrusts of the Precambrian Laurussian foreland. The thrusting occurred here during the Upper Visean and Lower Namurian continental collision between the two Paleozoic continents Laurussia and Gondwana.

A short seismic line is positioned in the Teplá-Barrandian area in West Bohemia. There are significant reflectors within Upper Proterozoic sedimentary-volcanic complexes. Some of them belong probably to tholeiitic basalt sills. Others may be interpreted as short Cadomian thrust faults between different Upper Proterozoic blocks. Other explanation could not be excluded taking in account lithological changes.

Introduction

The idea that the European Variscides (in Stille's terminology the northeast branch of the European Hercynides) resulted from a gigantic continental collision between "the old Africa and the old Europe" was originally presented by Franz Eduard Suess (1926). Specifically, it was suggested by Suess that large crystalline overthrust sheet movements (Wandertektonik) and following huge granitic plutonism (Intrusionstekto-nik) accompanied the collision process. Suess' description of the European Variscides reminds one to a great extent of the collision features in modern settings. After the half-centennial fixism, the European Variscides have recently been newly interpreted as the result of the Late Devonian and Carboniferous convergence and collisions of the European continent with smaller southern continental blocks and island arcs with pro-minent strike-slip faulting (e.g. Bambach et al. 1980, Dewey 1982, Behr 1983) etc., but the exact nature of the collision and accompanying crustal shortening is still obscure.

Several types of geophysical measurements and maps are available to aid in better under-standing of the deep structural features of the eastern part of the Bohemian Massif (easternmost European Variscides). A detailed gravity map with more than 4 stations per km^2 shows the gravity structures of the Bohemian Massif. The airborne geophysical survey covered practically the whole area of the Massif. It seems, however, to be general-ly accepted among the geophysicists community that the seismic reflection survey yields better results than other geophysical techniques in spatial resolution of deep geological structures. The deep crustal seismic reflection survey is the best technique currently available to study areas in which thrusting, either thin-skinned or thick-skinned, has occurred.

There are several reflection lines within the Rhenohercynian realm in western Europe. All of them show evident thrusting of the Rhenohercynian sedimentary sequences over their northern crystalline foreland. The most recent French Ecors line shows large thin-skinned thrusting of the southern wedge of metamorphic and sedimentary rocks over the London-Brabant Massif. The line in West Germany (Meissner et al. 1981) crossing the Aachen overthrust fault also shows the same style of thin-skinned thrusting; the line in South England (Chadwick et al. 1983) indicates evident Hercynian thrusting involving also the Precambrian basement. The aim of this paper is to contribute to a better know-ledge of structural geology in three important areas of the Bohemian Massif. Seismic reflection data, used in this study, show persuasively the thrust tectonics of the eastern termination of the European Variscides. Thrust tectonics is also observed within the older Cadomian Teplá-Barrandian zone.

Geologic Setting

The Bohemian Massif is built on its northeastern side by Culm flysch rocks (see Fig. 1) of Tournaisian and Visean age which are typical for the external units of the Hercynian fold belt in Europe. The line 3P/83 (Fig. 2) crosses the Hradec Greywackes (Upper Visean) defined by Patteisky (1928). The Hradec Greywackes, 800 m in the original sedimentary thickness, extend in NNE direction and are underlain by the Moravice Shales (Patteisky, 1928) of Middle Visean age and overlain by the Kyjovice beds of upper Visean-lower Namurian age (Kumpera, 1983). The line 5/83 (Fig. 2) was stretched through both the Hradec Greywackes and the Moravice Shales. Both profile directions are oblique to the main thrusting trend with an angle of about 50°.

Crustal shortening occurred immediately after the deposition of flysch and molasse sequences during the Upper Visean and Namurian A age, as it is evidenced by the undi-sturbed Karviná beds of Namurian B age overlying the Ostrava beds of Namurian A age.

The Variscan thrust and fold belt overthrust the Precambrian crystalline foreland — the Brunovistulicum (Dudek, 1980) which formed a part of the Laurussian North European platform. The Brunovistulicum is built of plutonic rocks of all kinds and by middle-grade metamorphites. The age of the plutonic and metamorphic rocks ranges mostly from 550 to 620 Ma (Dudek, 1980). The crystalline rocks of the Brunovistulicum are mostly covered by the Devonian and Dinantian sedimentary cover and beneath the Carpathians also by Upper Carboniferous sediments.

As regards the second area investigated, this paper deals with the old enigma of the over-thrust faulting of the Moldanubicum across the Moravicum in the southeastern part of the Bohemian Massif (Figs. 1 and 5). The thrusting hypothesis was suggested by Suess (1912) and for a long time was accepted by the majority of geologists working in this area (Kölbl, Preclik, Kettner, Zoubek, Kodym and others). Only 30 years ago, serious doubts about the Moldanubian overthrusting were raised. Our short line is situated within the variegated series of the Moravian Moldanubicum of the Náměšť-Moravský Krumlov area. Line 3/83 passes through the Gföhl gneisses, granulites and amphibolites and it is terminated 7 km south of the Náměšť fault (thrust fault postulated by Suess).

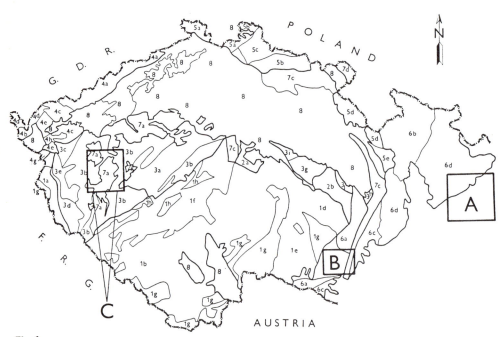

Fig. 1

Regional geological map of the Bohemian Massif in Czechoslovakia (slightly modified according to Mísař et al., 1983).

1 Moldanubicum (M.): **1a** M. of Český les, **1b** M. of Šumava (Böhmerwald), **1c** Czech M., **1d** Strážec M., **1e** Moravian M., **1f** Central Bohemian Pluton, **1g** Moldanubian Pluton, **1h** Islet region. 2 Kutná Hora-Svratka zone: **2a** Kutná Hora crystalline region (c.r.), **2b** Svratka c.r. 3 Central Bohemian zone: **3a** Barrandian Paleozoicum, **3b** Barrandian Proterozoicum, **3c** Teplá crystalline zone, **3d** Domažlice crystalline zone, **3e** West Bohemian Pluton, **3f** West Bohemian Mafic rocks, **3g** Železné Hory Pluton, **3h** Letovice c.r., **3i** Chrudim Proterozoic and Paleozoic zone, **3j** Polička c.r. 4 Krušné hory (Erzge-birge) zone: **4a** Krušné hory c.r., **4b** Smrčiny c.r., **4c** Krušné hory Pluton, **4d** Vogtland Paleozoic, **4e** Svatava c.r., **4g** Cheb-Dyleň c.r. 5 Lugicum: **5a** Lugic Pluton, **5b** Krkonoše-Jizera c.r., **5c** Krkonoše-Jizera Pluton, **5d** Orlice-Kladsko c.r., **5e** Zábřeh c.r. 6 Moravo-Silesian (Rhenohercynian) zone: **6a** Moravicum, **6b** Silesium, **6c** Brunovistulicum, **6d** Devonian and Lower Carboniferous. 7 Upper Pale-ozoic. 8 Post-permian units.

Area of investigation: **A** = Rhenohercynicum (Culm flysch), **B** = Moldanubicum, **C** = Teplá-Barrandian.

The third area of seismic investigation lies in the Cadomian deformed Proterozoic of the Teplá-Barrandian zone covered by undeformed Upper Paleozoic sediments. The underlying Proterozoic strata are composed of slightly metamorphosed deep-water sediments and island arc volcanites of Upper Proterozoic age (Kettner 1917, Cháb 1978). These rocks were strongly deformed during the Late Proterozoic Cadomian tectonic event and later on possibly also by the Variscan event.

Data Acquisition and Processing

Two other lines were shot using dynamite. Field parameters for lines 3/83 and 1/82 are listed below respectively: Source (single shot dynamite) 40 kg and 1.0 kg, shot depth 30 m and 20 m, shot interval 200 m and 20 m, geophone station interval 50 m and 10 m, recording 12 s and 3 s, channel splitspread 4750 m and 480 m, nominal fold of stack 12, sampling interval 2 ms. The Vibroseis technology was applied on two reflection lines in the Culm flysch area in 1983. The field techniques and processing employed by Geofyzika Brno to obtain the seismic reflection data examined here were essentially not modified from standard Vibroseis survey practices used in oil exploration. The compressional wave source in this survey consisted of three vibrators operating synchronously to transmit a sweep signal with frequency varying from 15 to 60 Hz. The duration of each sweep was 11 s with a total recording time of 3 s (3P/83) and 5 s (5/83) at sampling intervals of 2 ms. Vibrators were moved along 1/7 of VP spacing after each second sweep. The VP spacing was 25 m. Production profiling was executed in 12-fold (3P/83) and 24-fold (5/83) common reflection point mode.

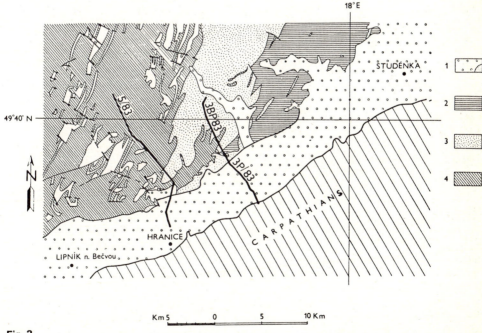

Fig. 2
Geological map of the Culm flysch area in northeast Moravia. **1** Neogene Carpathian foredeep, **2** Kyjovice layer, **3** Hradec layers, **4** Moravice layers.

The basic processing sequence for each profile was: demultiplex, correlation, sorting, elevation statics, velocity analysis, NMO corrections, mute, stacking, time-varying filtering, and AGC.

Geological Interpretation

The locations of the seismic profiles discussed in this paper are shown in Figs. 1, 2, 5 and 8. Two-way travel times are used throughout this paper. To convert the two-way times to approximate depths, multiply one-half the time by a velocity of 6 km/s for the Moldanubian and Brunovistulic basement rocks, for the compact Culm flysch rocks and also the West Bohemian Proterozoic sediments. Reflection dips are presumably high so that migration is needed and the position of some highly dipping reflectors is seriously drifted. Nevertheless, even non-migrated line should serve for our interpretation purposes showing a possible style of deformation rather than actual geological contacts.

The Rhenohercynian (Culm flysch area)

The Culm flysch Vibroseis lines 3P/83, 3BP/83 and 5/83 are not perpendicular to the main ESE trend of thrusting (Fig. 2). The lines are oblique in this respect with the angle of about $50°$. The consequences are that the angles and widths of thrusts are misrepresented. The unmigrated section of line 5/83 is shown in Fig. 3.

The most prominent feature of line 5/83 (Fig. 3) is a package of fairly steep, northwest-dipping wedge-shaped reflections in the northwestern part of the line. These reflections are probably due to a series of thrust faults within both rock formations, the Culm flysch and the Precambrian crystalline Brunovistulicum.

We interpret events A and A′ as the main Culm flysch thrust (floor thrust) lying on the hinterland dipping duplex (definition see in Boyer and Elliot, 1982) formed by horses of the Brunovistulic crystalline rocks with their Devonian and Dinantian sedimentary cover. Events B and C presumably represent the thrust surfaces within the imbricated duplex. Events D going practically to the surface are interpreted as an imbricate thrust system of Culm flysch rocks. The strong reflections E dipping to the SE belong entirely to the underlying autochtonous crystalline rocks of the Brunovistulicum and are probably not related to the Variscan collision tectonics. According to the surface geological mapping in the whole Culm area, all reflections A, B, C, and D probably come from mylonitized zones and are, therefore, not sedimentary in origin.

The 3P/83 line is migrated and converted into the depth model using known velocities of the Hradec Greywackes, the Badenian sediments (Fig. 4), and Devonian and crystalline rocks.

In the region of 3P/83 line (divided into three connected line 3BP/83, 3P/83), the Hradec layers of the Culm flysch are exposed in the northwest. Within the Culm flysch complex, a dipping band of reflectors can be clearly traced to a depth of about 3 km. The reflections are dipping at an angle of approximately $25°-30°$ to the northwest (Fig. 4) and are interpreted as a trace of the imbricated slices of the Hradec layers similarly like Moravice layers slices on line 5/83 which, according to the surface geological observations, are mylonitized.

The main band of the northwest dipping reflectors can be traced from the boundary of the Badenian depression (point 3.5, line 3P/83) to the northwest end of line 3BP/83 with an average dip of $25°-30°$ again. This band of strong reflectors observed within the depth interval 1.5 to 4.5 km is interpreted here as the eastern master decollement surface of the Variscan mountain front of the Bohemian Massif. The exact nature of the

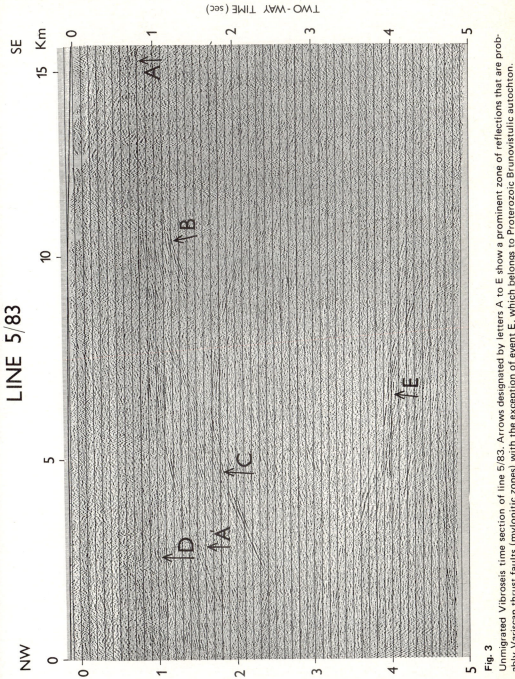

Fig. 3
Unmigrated Vibroseis time section of line 5/83. Arrows designated by letters A to E show a prominent zone of reflections that are probably Variscan thrust faults (mylonitic zones) with the exception of event E, which belongs to Proterozoic Brunovistulic autochton.

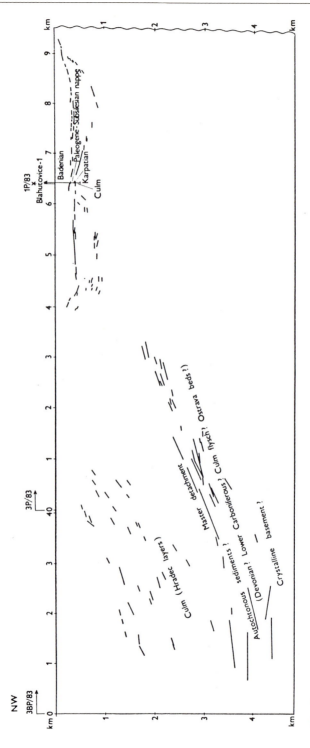

Fig. 4

Depth model of the Vibroseis line 3BP/83 and 3P/83. Note the Culm flysch imbrication above the basal master detachment.

reflections, however, is not clear. In our drawing, the sedimentary layers of the Devonian to Namurian A (Ostrava layers) are assumed to underlie the master thrust fault surface. But it cannot be excluded that a wide mylonitic zone exists there.

The Moldanubian Area

The length of the reflection profile in the Moldanubian area 3/83 is about 5 km (Fig. 5) in the southwest direction (Blížkovský et al., in press). In the time section (Fig. 6), numerous strong, coherent reflections can be easily identified throughout the range of travel times corresponding to the crust. Significant upper crustal subhorizontal reflectors are observed at 1.0−1.5 s and 1.9 s. These reflections probably do not correspond to the granulites, amphibolites and Gföhl gneisses near the surface, but run continuously beneath both complexes. They might be the original lower crust reflections of the planar structures originated in the course of plastic flow during high metamorphism or horizontally deposited magmatic rocks.

The most significant group of reflectors are reflections at time 1.9−2.5 s obliquely dipping to the south and horizontal reflections at times 3.0−3.5 s. The upper systems of dipping reflectors almost 2 km thick might represent the fault zone of the Molda-

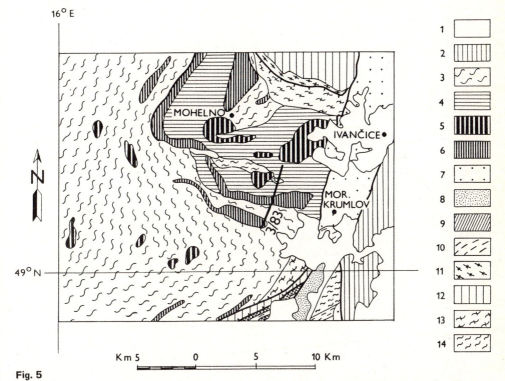

Fig. 5

Geological map of the Moldanubian area in southeastern Moravia (modified after J. Kalášek, 1963).

1 Neogene, **2** Brno Massif, **3** Gföhl gneisses, **4** granulites, **5** serpentinites, **6** amphibolites, **7** Permian of the Boskovice Furrow, **8** Culm flysch, **9** Moravian mica schists, **10** Moravian mica schists of the Miroslav horst, **11** Moravian outer phyllites, **12** Bíteš orthogneisses, **13** Moldanubian mica schists, **14** crystalline rocks of the Brno Massif within the Miroslav horst.

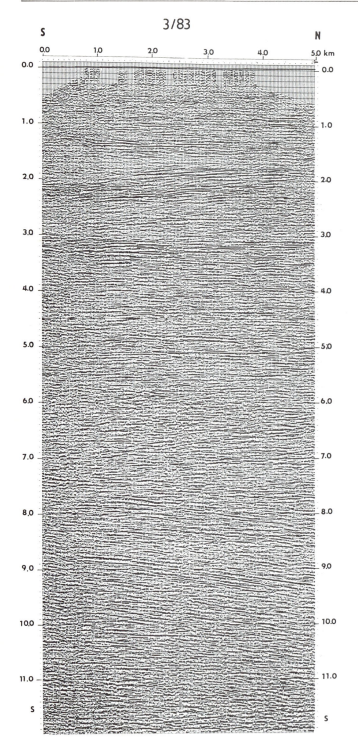

Fig. 6
Unmigrated time section
of line 3/83.

nubian overthrusting. The reflections are presumably due to the mylonitized strips of rocks that should be at least 80 m thick each to become a distinct reflector (Fountain et al., 1984). On the depth model from the first three seconds TWT of time section (Fig. 7) where all seismic events are located at the level of the profile, it can be seen the boundaries are dipping to the outcrop of the Náměšť fault. These results are in accordance with the Suess' conception.

The thrusting of the Moravicum over Brunovistulicum interpreted by Jaroš and Mísař (1974) might be indicated by lower subhorizontal reflections between 3.0 and 3.5 s.

The rest of the upper crust and the middle part of the lower crust between approximately 4.0 and 7.0 s are relatively poor in reflections. Below 7.0 s there are again relatively strong, coherent reflections or groups of reflections at lower crustal levels. Practically all of them are gently dipping to the north. The majority of lower crust reflections on profile 3/83 are between 8.8−9.4 s which appear after depth transformation several km south of the beginning of the profile. Thus they evidence the complicated structure of the lower crust at depths 25−30 km beneath the eastern margin of Moldanubicum. Explanation for such a layered character includes igneous layering, tectonic layering or metasedimental origin. Less distinct, short and discontinuous horizontal reflections between 11 and 12 s, namely in the lowest part of the section, presumably reflect the transition zone crust-mantle which is in accordance with DSS results from this area (Mayerová et al., 1985).

The Teplá-Barrandian

The third line 1/82, which is 6 km long, is situated in the Teplá-Barrandian area which is built up here (west of Plzeň) of Upper Proterozoic metasediments and volcanites covered by Upper Carboniferous and Permian post-Variscan sediments with coal seams (Fig. 8). Although the data acquisition was employed for the purposes of coal prospecting, we obtained some interesting seismic events from the Upper Proterozoic strata underlying the Late Paleozoic sediments (Fig. 9, time 0.5−2.9 s). Inside the zone of Proterozoic rocks, two relatively continuous and layered bands of reflection with opposite dip can be traced in the time section (Fig. 9). One band runs through the whole section from 1.5 s to 0.9 s and the other one from 2.7 s to 1.4 s. On the depth model (Fig. 10), the strong reflectors steeply dipping to the SE are either side effects outside from the plane of the profile or these events can correlate with known surface geological features. In that case they correspond to Late Proterozoic island arcs tholeiites (spilites) which crop out here exactly in the prolongation of the reflectors. The significant group of layered reflections dipping to the NW at a depth of 1.5−4.0 km can be interpreted as a planes of intra-Proterozoic thrust faults although lithological explanation cannot be excluded.

Conclusions

Seismic reflection profiling has yielded important information from the easternmost part of the European Variscides. The four reflection lines from the Bohemian Massif have brought new insight into this classical geological terrain. The main results are as follows:

1. The seismic results from the eastern margin of the Bohemian Massif imply that WNW continental underthrusting of the Brunovistulic continent (part of Laurussia) took place in the European Eastern Variscides during Upper Visean and Lower Namurian ages.

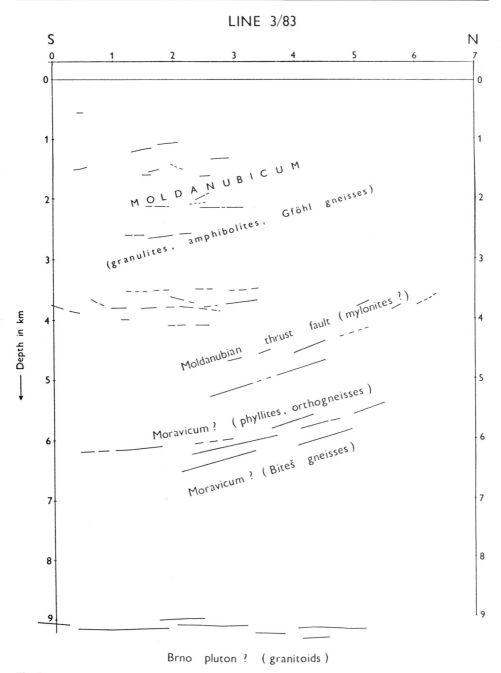

Fig. 7
Depth model of line 3/83 with geological interpretation.

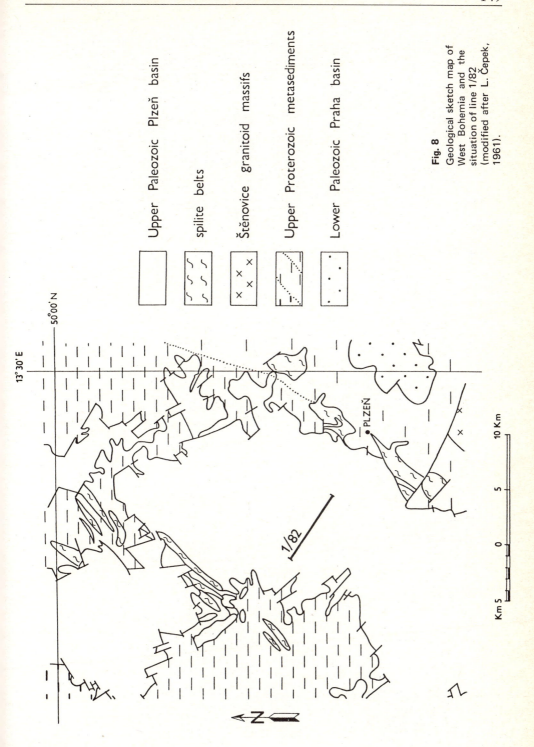

Upper Paleozoic Plzeň basin

spilite belts

Štěnovice granitoid massifs

Upper Proterozoic metasediments

Lower Paleozoic Praha basin

Fig. 8
Geological sketch map of West Bohemia and the situation of line 1/82 (modified after L. Čepek, 1961).

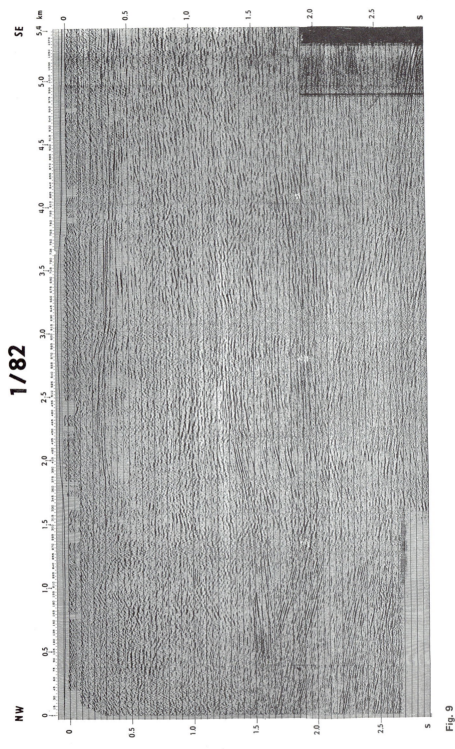

Fig. 9
Unmigrated time section of line 1/82.

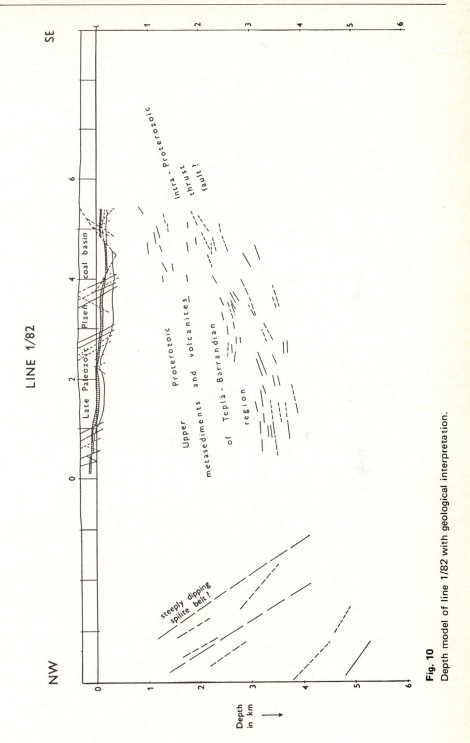

Fig. 10
Depth model of line 1/82 with geological interpretation.

2. The Culm flysch rocks of Visean age are developed into a typical "schuppen" system, whereas the underlying basement thrust sheet of Brunovistulic rocks probably forms a duplex structure. The thin-skinned style of thrusting is similar to that of West Germany and northern France.
3. The Moldanubian reflection line 3/83 yields an extremely complex picture of seismic events on all crustal levels. Upper crustal reflections seem to correspond to Variscan thrust faulting of the Moldanubicum over Moravicum and Moravicum over Bruno-vistulicum.
4. Cadomian terrain in western Bohemian seem to be also significantly influenced by thrust faulting of Late Proterozoic period.

References

Bambach, R. K., Ch. R. Scotese, and A. M. Ziegler (1980): Before Pangea: The Geographies of the Paleozoic World. American Scientist, 68, 26–38.
Behr, H. J., W. Engel, and W. Franke (1982): Variscan wildflysch and nappe tectonics in the Saxo-thuringian Zone (northeast Bavaria, West Germany). American Journal of Science, 282, 1438–1470.
Blížkovský, M., A. Dudek, I. Ibrmajer, Z. Mísař, M. Suk, and Č. Tomek (1985): A new deep seismic reflection line in the Moldanubicum. Geologický průzkum (in press), (in Czech).
Boyer, S. E. and D. Elliot (1982): Thrust systems. The American Association of Petroleum Geologists Bulletin, 66, 1196–1230.
Čepek, L. (editor) (1961): Geological map of Czechoslovakia – sheet Plzeň, Geological Survey of Czechoslovakia, Prague.
Cháb, J. (1978): Aproposal of lithostratigraphic and lithologic terminology for upper Proterozoikum of the Teplá-Barrandian area (in Czech). Věst. Ústř. Úst. geol., Praha, 53, 43–60.
Chadwick, R. A., N. Kenolty, and A. Whittaker (1983): Crustal structure beneath southern England from deep seismic reflection profiles. Journal of the Geological Society, London, 140, 893–911.
Dewey, J. F. (1982): Plate tectonics and the evolution of the British Isles. Journal of the geological Society, London, 139, 371–412.
Dudek, A. (1980): The crystalline basement block of the outer Carpathians in Moravia: Brunovistuli-cum. Rozpravy Čes. Akad. věd, roč. 90 – sešit 8, Praha.
Fountain, D. M., Ch. A. Hurich, and S. B. Smithson (1984): Seismic reflectivity of mylonite zones of the crust. Geology, 12, 125–198.
Jaroš, J. and Z. Mísař (1974): Beckenbau der Svratka-Kuppel und seine Bedeutung für das geodyna-mische Modell der Böhmischen Masse. Sbor. geol. Věd, Geol., Praha, 26, 29–79.
Kalášek, J. (editor) (1963): Geological map of Czechoslovakia-sheet Brno, Geological Survey of Czechoslovakia, Prague.
Kettner, R. (1917): Versuch einer stratigraphischen Einteilung des böhmischen Algonkiums, Leipzig. Geol. Rdsch., 8, 169–188.
Kumpera, O. (1983): Geology of the Lower Carboniferous of the Jeseníky block (in Czech), Aka-demia Prague. 172 pp.
Mayerová, M., Z. Nakládalová, I. Ibrmajer, and H. Herrmann (1985): The three-dimensional distribu-tion of Moho in Czechoslovakia. 8. Geophysical congress of cz. geophysicists, České Budějovice 1985 (in Czech).
Meissner, R., J. Bartelsen, and H. Murrawski (1981): Thin-skinned tectonics in the northern Rhenish Massif, Germany. Nature, 290, 399–401.
Mísař, Z., A. Dudek, V. Havlena, and J. Weiss (1983): Geology of Czechoslovakia – I. part – Bohe-mia Massif, SPN Prague (in Czech).
Patteisky, K. (1919): Die Geologie und Fossilführung der mährisch-schlesischen Dachschiefer- und Grauwackenformation, Opava. 125 pp.
Suess, F. E. (1912): Die Moravischen Fenster und ihre Beziehung zum Grundgebirge des Hohen Ge-senkes. Denksch. Akad. Wiss. Wien, 88, 541–631.
Suess, F. E. (1926): Intrusionstektonik und Wandertektonik im variszischen Grundgebirge. Verlag von Gebrüder Borntraeger, Berlin, 268 pp.

On the Volcanology of the West Eifel Maars

V. Lorenz

Institut für Geowissenschaften, Universität Mainz, Saarstr. 21, D-6500 Mainz, Federal Republic of Germany

The Quaternary alkali-basaltic volcanic field of the West Eifel is the classic maar region of the world. Related to the formation of the continental rift zone which extends through Central Europe, ultrabasic magmas rose from the upper mantle through the continental crust and fed 240 small volcanoes. Next to scoria cones with their lava flows, maars are most frequent and represent 25 % of the volcanoes. Since 1820 and until 1970 their origin was mostly believed to be related to explosive exsolution of juvenile volatile phases. Since 1970, however, several authors have suggested that the West Eifel maars are phreatomagmatic in origin, i.e. rising magma contacted groundwater in near-surface levels, thus causing water vapor explosions.

The maars are large craters several hundred m to 1700 m in diameter and up to 200 m in depth. They are cut into the preeruption country-rocks which are Devonian and Mesozoic in age. With the exception of two maars, all others are located within preexisting valleys. The craters are surrounded by a rim of numerous thin pyroclastic beds. This clearly shows that formation of the large craters is the result of numerous individual eruptions which closely followed each other and were not able to excavate the craters explosively.

The main characteristics of the ejecta beds are the vesicle-poor to vesicle-free nature of their juvenile ash grains, lapilli and bombs, and their high content in country-rock clasts. The former shows that exsolution of juvenile volatile phases did not play a major role in near-surface levels and, therefore, cannot have been responsible for the near-surface explosive activity of the maars. In contrast, it suggests that an external agent, i.e. groundwater, was responsible for the particular explosive eruptive behavior of the maars. The latter fact points to a high energy explosive process at the near-surface level of underground explosive activity. Other volcanologically important data are indicative of eruption-clouds — so-called base surges — which again and again flowed radially outward along the ground surface surrounding the craters and deposited the numerous individual pyroclastic beds. Some are also indicative of wet or moist eruption deposits requiring large amounts of condensating steam to have participated in the eruptions. The pyroclastic deposits from historic, observed phreatomagmatic eruptions, such as Surtsey/Iceland 1963—1966, Taal/Philippines 1965, and the Ukinrek Maars/Alaska 1977, show some or all of these characteristics and, therefore, support the model of phreatomagmatic formation of the Quaternary unobserved West Eifel maar eruptions.

Ejection of large quantities of fragmented country-rocks by the underground explosive activity results in a syneruptive mass deficiency at depth and repeated collapse of the wall-rocks. This collapse is propagated upward and a collapse crater, a maar, forms at the surface cut into the preeruption country-rocks. As the rim deposits of the Eifel maars only rarely contain non-phreatomagmatic deposits, i.e. scoria beds, it is obvious that groundwater had access to the rising magma nearly continuously during the whole eruptive activity of the respective maars.

It has been already stated above that nearly all West Eifel maars are located in valleys. This coincidence, also found in many other maar regions of the world, points to hydraulically rather active zones of structural weakness beneath the valley floors. These zones of structural weakness are used by groundwater and rising magma, and erosion has relatively easily shaped them into valleys. Thus the coincidence of maars being located in valleys clearly shows what has been known for long: that valleys are frequently underlain by zones of structural weakness, that zones of structural weakness are frequently hydraulically active, and that magma during its rise towards the surface frequently makes use of preexisting zones of structural weakness because of the prevailing stress-field conditions. Therefore, it should not be astonishing that maars are frequently located in valleys, i.e. on hydrogeological favorable sites.

The scoria cones of the West Eifel volcanic field are located off the valleys. They formed through lava fountaining because of exsolution of the juvenile volatile phases. Many of these scoria cones, however, are nested in small maars and they are therefore called scoria cones with an initial maar. At these volcanoes phreatomagmatic eruptions caused formation of a small maar in a first phase, and then a scoria cone formed in a second phase within the initial maar because of lack of groundwater but a continuous rise of magma. At many scoria cones, additional short or longer phreatomagmatic phases interrupted the lava fountaining.

There are thus three different types of alkali-basaltic volcanoes in the West Eifel: 1. scoria cones with or without phreatomagmatic phases during the lava fountaining activity; 2. scoria cones with an initial maar with or without further phreatomagmatic phases during the lava fountaining activity; 3. maars. These different volcano types clearly point to differing hydrogeological conditions at the site of magma rise, influencing eruptive style and duration of phreatomagmatic eruptions or lava fountaining.

References

Büchel, G. (1984): Die Maare im Vulkanfeld der Westeifel, ihr geophysikalischer Nachweis, ihr Alter und ihre Beziehung zur Tektonik der Erdkruste. Dr. rer. nat. thesis (unpubl.), Joh. Gutenberg-Universität, Mainz, 385 pp.

Büchel, G. and Lorenz, V. (1982): Zum Alter des Maarvulkanismus der Westeifel. N. Jb. Geol. Paläont., Abh., 163, 1–22.

Büchel, G. and Mertes, H. (1982): Die Eruptionszentren des Westeifeler Vulkanfeldes. Z. dt. geol. Ges., 133, 409–429.

Kienle, J., Kyle, P., Self, S., Motyka, R. J., and Lorenz, V. (1980): Ukinrek maars, Alaska, I. April 1977 eruption sequence, petrology and tectonic setting. J. Volcanol. Geotherm. Res., 7, 11–37.

Lorenz, V. (1984): Explosive volcanism of the West Eifel volcanic field/Germany. In: Kornprobst, J. (ed.): Kimberlites. I: Kimberlites and related rocks, 299–307, Elsevier, Amsterdam.

Lorenz, V. and Büchel, G. (1980): Zur Vulkanologie der Maare und Schlackenkegel der Westeifel. Mitt. Pollichia, 68, 29–100.

Meyer, M. (1985): Zur Entstehung der Maare in der Westeifel. Z. dt. geol. Ges., 136, 141–155.

Schmincke, H.-U., Lorenz, V., and Seck, H. A. (1983): The Quarternary Eifel volcanic fields. In: Fuchs, K., Gehlen, K. von., Mälzer, H., Murawski, and Semmel, A.: Plateau uplift. The Rhenish Shield — a case history, 139–151, Springer, Berlin, Heidelberg, New York, Tokyo.

Seismicity and Seismotectonics of the Rhenish Massif, Central Europe

L. Ahorner

Geological Institute, University of Cologne, Zülpicher Str. 49, D-5000 Köln 1

Extended Abstract

The contribution gives a review of the earthquake activity of the Rhenish Massif, a Hercynian folded region which has undergone some 300 m of plateau uplift since the Late Tertiary. Seismotectonic implications on geodynamic processes controlling crustal uplift are discussed. The study is based on a detailed investigation of the historical seismicity since 1500 A.D., and on a special microearthquake survey carried out during the period 1976–1982 within the scope of the multidisciplinary DFG-project "Rhenish Massif". A local station network consisting of 11 high sensitivity analog or digital seismographs has been installed in the Rhenish Massif by the Department of Earthquake Geology of the University of Cologne in cooperation with the Geophysical Institutes of the Universities of Frankfurt and Karlsruhe. More than 800 local earthquakes of tectonic origin, ranging in magnitude from $ML = 0$ to 4.7, were recorded by the network during the 6-year study. Most of them could be located with reasonable accuracy (location error smaller than 1 km horizontally and 2 km vertically). Fault plane solutions were derived, in general, for shocks with magnitudes greater than $ML = 3$.

The main features of the historical seismicity pattern are reproduced by the short term microseismicity, although differences exist in detail (see Fig. 1). Focal regions with large historical earthquakes show, for instance, in general an unusual low level of microseismicity. From S-wave displacement spectra of digitally recorded earthquakes, the seismotectonic source characteristics (seismic moment, source radius, average dislocation, and stress drop) have been determined using the Brune (1970) source model. Stress drops found vary between several and more than 150 bars. Small shocks have in general smaller stress drops than larger ones. A local anomaly of the stress drop regime correlating with an anomaly of the focal depth distribution is probably existing in the central part of the Rhenish Massif near the Neuwied Basin and the Late-Quaternary volcanic field of the Laacher See.

The overall pattern of seismicity within the Rhenish Massif and its vicinity is clearly related to stress-field controlled block movements along major crustal fracture zones, like the Rhine Graben system, and shows no direct relationship to the massif uplift. Only in some special regions, e.g. the Hohes Venn in the northwestern part of the massif, where an exceptional large present-day uplift rate of 1.6 mm/y has been found by geodetical means, some evidence for a causal correlation between seismicity and plateau uplift is achieved.

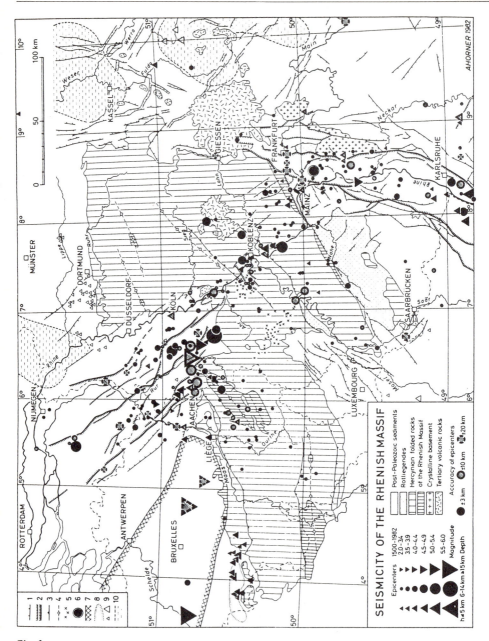

Fig. 1

Historical seismicity pattern and geological features of the Rhenish Massif. 1 fault zone active during Tertiary time; 2 fault zone active until Quaternary and Recent time; 3 thrust fault of Hercynian age; 4 major Hercynian anticline; 5 Quaternary volcano; 6 earthquake swarm; 7 border line of the Caledonian Brabant massif; 8 Hainaut basin; 9 seismic events due to mining activity; 10 border line of underground salt deposits.

The seismotectonic regime of the Middle Rhine zone between Mainz and Bonn is characterized by extensional rifting along a broad NW-SE trending fracture zone, which follows approximately the course of the river Rhine and connects the major seismoactive features of the Upper Rhine Graben with those of the Lower Rhine Embayment. The Rhenish Massif is bisected by this zone into two half parts which were shifted away from each other in SW-NE direction. From the geological point of view, the hidden zone of active rifting consists most probably — even in the deeper underground — not of large pervading fault lines, but of numerous smaller faults orientated sub-parallel or in en-echelon configuration. This implies a rather diffuse picture of the ruptural deformation of the crust which might be caused by the slaty composition of the majority of rocks in the upper part of the crust.

References

Ahorner, L. (1983): Historical seismicity and present-day microearthquake activity of the Rhenish Massif, Central Europe. In: K. Fuchs et al. (eds.), Plateau Uplift, p. 198—221; Springer, Heidelberg.

Ahorner, L., Baier, B., Bonjer, K.-P. (1983): General pattern of seismotectonic dislocation and the earthquake-generating stress field in Central Europe between the Alps and the North Sea. In: K. Fuchs et al. (eds.), Plateau Uplift, p. 187—197; Springer, Heidelberg.

Geophysics in Structural Research on the Rhenish Massif

A. Vogel

Institut für Geophysikalische Wissenschaften, Freie Universität Berlin, Podbielskieallee 60, D-1000 Berlin 33

Extended Abstract

Results of geophysical studies are summarized, which have contributed to elucidate the structure, origin, evolution and present evolutionary processes of the Rhenish Massif. The evaluation of evidence from the early evolution of the Rhenish Massif leads to a simplified model of the original geodynamic processes.

The depositional phases indicate that the basin formation and sedimentation began in the southern part of the Rhenish Massif in the early Devonian and propagated towards north-west through the Devonian to the early Carboniferous (Franke et al. 1978). Studies on metamorphism by the K/Ar method show that the deformational phase of the Hercynian orogeny began in the south-eastern part of the Massif 330 Ma ago and migrated to the north-western part 300 Ma ago (Ahrend et al. 1978; Weber 1981).

Obviously the Rhenish Massif owes its origin to mantle convection that caused a dilational wave followed by a compressional wave both propagating from SE to NW causing stress fields with main axis in the direction of propagation. Crustal thinning and basin formation with sedimentation progressing from SE to NW was followed by lateral compression with crustal thickening, folding, faulting and thrusting.

The Mesozoic is characterized by times of uplift and subsidence changing in different parts of the Rhenish Massif (Murawski et al. 1983). During the Tertiary and Quaternary, the evolution of the Rhenish Massif was greatly influenced by the geodynamic processes which caused the collision of the African and European plate, with Alpine orogeny as the most spectacular phenomenon. In the foreland of the Alpine collision front a stress domain developed with a main compressional axis in SSE-NNW direction. The Middle-European crust reacted with folding and fracture formation.

The most predominant fracture zone which was formed extends from the Upper Rhine Valley across the Rhenish Massif into the Lower Rhine Embayment. Focal mechanisms of earthquakes, which occur in this fracture zone, indicate compressional stress in SSE-NNW direction and dilatation along a perpendicular main stress axis (Ahorner et al. 1983).

Since the Middle Tertiary the Rhenish Massif has been uplifted by 150 m and more. Dating and correlation of the terraces of the Rhine valley and other valleys indicate that uplift has not been uniform for the entire Massif and that uplift rates have been rather episodic (Fuchs et al. 1983).

Main uplift with 1.6 mm/year occurs today in the Hohes Venn while in other parts vertical movements are very low and of different sign (Mälzer et al. 1981). Reflection seismic studies and earthquake activity in this area indicate that uplift is caused by thrusting along the reactivated Aachen thrust fault of Hercynian age (R. Meisner et al. 1983).

Seismic and magnetotelluric studies in other parts of the Massif have revealed deep crustal structures which are dipping toward SSE (Giese 1983; Jödicke et al. 1983) in accordance with the geodynamic model developed by Weber (1978, 1981) and Behr (1978). It seems that the present stress field provides favorable conditions for the reactivation of Hercynian thrust zones, which appear to be the cause for Late Tertiary to present uplift.

Velocity jumps corresponding to the Conrad discontinuity are of variable horizontal extent and depth. The results of deep seismic soundings indicate that the conditions at the core-mantle boundary vary considerably in various parts of the Massif. Based on the seismic velocity structure, a density distribution could be found which fits the main features of the observed Bouguer field. Most striking is the fact that seismic velocities and Bouguer field in the eastern part of the Massif are distinctly higher than in the western part (Drisler and Jacoby 1983).

Teleseismic studies indicate a low velocity zone in 50−200 km depth centered beneath the Quaternary volcanic district of the West-Eifel (Raikes and Bonjer 1983). This low velocity zone, which however could not be identified by gravity and magnetotellurics, is interpreted as a zone of partial melting. The geodynamic processes which caused compressional forces associated with land uplift obviously produced partial melting in the upper mantle. Preexisting tensional fracture zones have provided the opportunity for the lavas to ascend to the surface causing the widespread and spectacular volcanism of the Westerwald and Eifel in Tertiary to Quaternary times which extinguished only less than 10 000 years ago.

References

Ahrendt, H., J.-C. Hunziker and K. Weber, 1978: K/Ar-Altersbestimmungen an schwach metamorphen Gesteinen des Rheinischen Schiefergebirges. Z. dt. geol. Ges. 129, 229−249.

Ahorner, L., B. Baier and K.-P. Bonjer, 1983: General Pattern of Seismotectonic Dislocation and the Earthquake-Generating Stress Field in Central Europe Between the Alps and the North Sea. In Plateau Uplift, The Rhenish Shield − A Case History, Eds. K. Fuchs et al., Springer Verlag, Berlin−Heidelberg−New York−Tokyo.

Behr, H.-J., 1978: Subfluenzprozesse im Grundgebirgsstockwerk Mitteleuropa. Z. Dtsch. Geol. Ges. 129, 283−318.

Drisler, J. and W. R. Jacoby, 1983: Gravity Anomaly and Density Distribution of the Rhenish Massif. In Plateau Uplift, The Rhenish Shield − A Case History, Eds. K. Fuchs et al., Springer Verlag, Berlin−Heidelberg−New York−Tokyo.

Franke, W., W. Eder, W. Engel and F. Langenstrassen, 1978: Main aspects of geosyncinal sedimentation in the Rhenohercynian Zone. Z. dt. geol. Ges. 129, 201−216.

Fuchs, K., K. von Gehlen, H. Mälzer, H. Murawski und A. Semmel (Editors), 1983: Plateau Uplift, The Rhenish Shield-A Case History. Springer Verlag Berlin−Heidelberg−New York−Tokyo.

Giese, P., 1983: The Evolution of the Hercynian Crust − Some Implications to the Uplift Problem of the Rhenish Massif. In Plateau Uplift, The Rhenish Shield − A Case History, Eds. K. Fuchs et al., Springer Verlag, Berlin−Heidelberg−New York−Tokyo.

Jödicke, H., H. Untiedt, W. Olgemann, L. Schulte and V. Wagenitz, 1983: Electrical Conductivity Structure of the Crust and Upper Mantle Beneath the Rhenish Massif. In Plateau Uplift, The Rhenish Shield − A Case History, Eds. K. Fuchs et al., Springer Verlag, Berlin−Heidelberg−New York−Tokyo.

Mälzer, H., G. Schmitt and K. Zippelt, 1981: Recent vertical movements and their determination in the Rhenish Massif. Tectonophysics, 52, 167−176.

Meisner, R., M. Springer, H. Murawski, H. Bartelsen, E. H. Flüh and H. Dürschner, 1983: Combined Seismic Reflection-Refraction Investigations in the Rhenish Massif and Their Relation to Tectonic Movements. In Plateau Uplift, The Rhenish Shield — A Case History, Eds. K. Fuchs et al., Springer Verlag, Berlin—Heidelberg—New York—Tokyo.

Murawski, H., H. J. Albers, P. Bender, H.-P. Berners, St. Dürr, R. Huckriede, G. Kauffmann, G. Kowalczyk, P. Meiburg, R. Müller, A. Muller, S. Ritzkowski, K. Schwab, A. Semmel, K. Stapf, R. Walter, K.-P. Winter and H. Zankl, 1983: Regional Tectonic Setting and Geological Structure of the Rhenish Massif. In Plateau Uplift, The Rhenish Shield — A Case History, Eds. K. Fuchs et al., Springer-Verlag, Berlin—Heidelberg—New York—Tokyo.

Raikes, S. and K.-P. Bonjer, 1983: Large Scale Mantle Heterogenity Beneath the Rhenish Massif and Its Vicinity from Teleseismic P-Residuals Measurements. In Plateau Uplift, The Rhenish Shield — A Case History, Eds. K. Fuchs et al., Springer Verlag, Berlin—Heidelberg—New York—Tokyo.

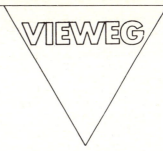

Rodney A. Gayer (Ed.)

The Tectonic Evolution of the Caledonide-Appalachian Orogen

1985. VI, 194 pages, 45 fig. 16 x 24 cm. (Earth Evolution Sciences. International Monograph Series on Interdisciplinary Earth Science Research and Applications, ed. by Andreas Vogel.) Softcover

The geology of the Caledonide-Appalachian orogen is probably the most intensively studied of all mountain chains, and yet its origins and evolution are still highly controversial. Interest in the orogen has been heightened in recent years by a vast amount of new information arising from work in connection with I. G. C. P. project no. 27 – the Caledonide Orogen. It is clear that, in addition to modelling the Caledonide-Appalachian orogen on present plate boundaries, it is necessary to recognise the importance of major fault contacts. As with the Western Cordillera of North America, where stratigraphic, structural and palaeomagnetic studies have demonstrated the presence of a large number of suspect terranes, juxtaposed by major strike-slip displacements, so the Caledonide-Appalachian orogen is best investigated by terrane analysis. Each terrane is regarded as fault displaced unless a definite connection between adjacent terranes can be established.

The articles presented in this thematic issue have been carefully selected to give a wide coverage of the orogen, reviewing the important facets of its evolution in terms of plate tectonics and suspect terrane analysis. Such an approach, of necessity, covers a wide spectrum of earth science disciplines but every effort has been made to integrate the individual articles into a general framework. Thus the first article on British Caledonian terranes lays the foundations for the main reviews and the final article presents a coordinated tectonic model for the evolution of the belt, integrating the material from the individual contributions. The account presented in this issue represent an up to date overview of the evolution of one of the most intriguing of orogenic belts.

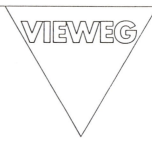

Andreas Vogel (Ed.)

Terrestrial and Space Techniques in Earthquake Prediction Research

1979. VIII, 712 pages, 265 fig. 17 x 24,5 cm. (Progress in Earthquake Prediction Research, Vol.1; ed. by Andreas Vogel.) Hardcover

The tremendous progress in earthquake dynamic research, the development of sophisticated technologies, the advent of the space age, and the call for reduction of earthquake hazards and for disaster prevention programs have spurred interdisciplinary research to discover the earth-qake generating processes. This book presents forty-nine papers given at an international workshop on Monitoring Crustal Dynamics in Earthquake Zones, held in Straßbourg, France, August 29 – September 5, 1979.

The first chapter introduces the measuring techniques for monitoring crustal dynamics in earthquake zones. Following chapters contain specialist papers in the following areas: micro-earthquake networks; continous recording of crustal deformation; geodetic ground techniques and gravity surveys; temporal variation of geophysical rock properties; combination of terrestrial techniques; and space techniques. The last chapter presents the concept of earthquake dynamics control networks and modelling procedures for the interpretation of the complexity of observations.

A. Mete Isikara, Andreas Vogel (Eds.)

Multidisciplinary Approach to Earthquake Prediction

1982. XVIII, 578 pages, 246 fig. 16,5 x 24,5 cm. (Progress in Earthquake Prediction Research, Vol. 2; ed. by Andreas Vogel.) Hardcover

Studies of the sources of earthquakes, the only way towards a successful prediction requires a broad interdisciplinary approach and cooperation between scientists in various fields of earth sciences. Seismotectonic studies, observation of strain accumulation and investigation of the physical state and temporal changes of rock properties in the earthquake source region are the main lines to be followed. Modelling of earthquake generating processes constrained by the complexity of precursory phenomena is a necessary basis of prediction research.

The North-Anatolian fault zone is an area of extremely high earthquake risk. Large and destructive earthquakes are most likely to occur in a most populated and industrialized part of Turkey. The chances "to capture an earthquake" by proper selection of observation sites and networks and to find criteria for earthquake prediction by observation of earthquake generating processes seem rather promising in the case of the North-Anatoiian fault zone.

An international and interdisciplinary conference was held in Istanbul. Experts in the fields of geophysics, geology and geodesy from many countries in Europe and overseas concerned with earthquake research and hazard assessment took part. They met to present results of earlier research, to discuss current research activities relevant to earthquake e prediction in the North-Anatolian fault zone and to work out and organize a program for earthquake prediction research in the mist riskful part of the fault zone.